Trends in Information Transfer

Trends in Information Transfer

Edited by

Philip J. Hills

Greenwood Press, Westport, Connecticut

Library of Congress Cataloging in Publication Data
Main entry under title:

Trends in information transfer.

Includes index.
Contents: Information technology and the printed word / Robert Campbell -- The importance of information in information technology / Jon Maslin -- Redesigning journal articles for on-line viewing / Maurice B. Line -- [etc.]
1. Information science--Addresses, essays, lectures.
2. Printing, Practical--Addresses, essays, lectures.
I. Hills, P. J. (Philip James)
Z665.T92 020'.28'54 82-3021
ISBN 0-313-23600-3 (lib. bdg.) AACR2

First published 1982

Published in the United States and Canada by
Greenwood Press, a division of Congressional
Information Service, Inc., Westport, Connecticut

English language edition, except the United States and Canada, published by Frances Pinter (Publishers) Limited

Library of Congress Catalog Card Number: 82-3021

ISBN: 0-313-23600-3

Printed in the United States of America

10 9 8 7 6 5 4 3 2 1

CONTENTS

PREFACE

This book arose from the response to my first book in this area, *The Future of the Printed Word* (Frances Pinter, 1980 and Greenwood Press 1980). In the preface to that book I suggested that the papers in it were like those at a conference and that a second volume could act like the discussion sessions at a conference – amplifying and extending the coverage of the subject. Comments, suggestions and papers were forthcoming, and the present collection of acticles was assembled – all by specialists in the field of publishing, librarianship, information science, computing and education. It is appropriate that it is published in 1982, Information Technology Year in the UK and World Communication Year to follow in 1983, which heralds an increasing awareness and debate in this important area. This book is offered as part of that debate.

Chapter 1 gives an introductory article by Bob Campbell and is followed by Jon Maslin emphasising at the outset the importance of information in information technology. Maurice Line contributes an article on redesigning journal articles for online viewing and this links with papers by Philip Hills and Paul Lefrere which are concerned with future possibilities in the preparation and use of information. John Michel Gibb reviews information transfer in Europe in Chapter 4 and this is followed by three chapters on different methods of information output, videotex, microform and high quality computer printing.

Chapter 10 by Geoffrey Hamilton contains a select bibliography on libraries and the new technology. This was written in response to a comment that in the last volume aspects relating to libraries and librarians had been neglected. The book ends with a survey of publishers' opinions on new technology and information transfer.

My thanks to all the contributors who so nobly held to the tight deadlines for the production of the final copy and my thanks also to all those who sent in suggestions and contributed ideas which led to the final shaping of this book.

Thanks are due to the British Library for permission to reproduce an updated version of the select bibliography 'Libraries and the New Technology'. This was originally issued by the Library Association Library in its series of Reading Lists.

Philip J. Hills
Leicester
January, 1982

1. INFORMATION TECHNOLOGY AND THE PRINTED WORD

ROBERT CAMPBELL
Department of Education,
University of York, UK

Microelectronics is perhaps the most influential technology of the century. It is cheap and becoming cheaper yet is exceptionally reliable. It is rapidly becoming technologically more advanced, resulting in the availability of small, low energy consuming devices capable of performing complex operations. It can extend and even replace a wide range of intellectual and intuitive skills. However it is not the technology itself but the specific applications of the technology which present us with the challenge. We have the opportunity to improve and develop our existing products, to modify our production processes, to introduce new production techniques and new products. Information technology which combines the technologies of computing and telecommunications is one particularly important application of microelectronics. This offers many novel and exciting ways of gathering, manipulating and presenting information. One production process and product which this is set to influence is printing and the printed word.

The production of print, whether it be book, newspaper, journal, letter or invoice requires that information be handled. Such information handling requires a sequence of five stages. These are: input, processing, storage, transmission and output. The new information technology is poised to influence all of these.

There is the danger that microelectronics and computing are seen as the only technologies which are making significant contributions to the development of information technology. True, these are vastly influential technologies but their importance must not overshadow the fundamental role of telecommunications and, in particular, the revolutionary nature of current developments in fibre optics. Communication systems based on fibre optic technology utilise thin wires of silica to guide light waves rather than copper wires to carry electrical signals as in the conventional systems.

A fibre optic cable packs a large information carrying capacity into a very small cable volume and provides a faster, wider, safer and cheaper service than copper cables can. There is the additional advantage that the signals which are carried are free from electrical and cross-channel interference. The signals are sent in digitalised form by the use of high speed switching lasers, enabling many thousands of channels to be used simultaneously for the transmission of coded data or conversation over thousands of kilometres.

British Telecom intend to install some ten thousand kilometres of optical fibre cable during the 1980s to provide a high quality trunk network connecting the major cities. Several other countries have similar plans. There is also an intention to provide a fibre optic link across the Atlantic. This submarine cable would be of three strands and designed to be capable of carrying some twelve thousand calls at one time. It is interesting to note that regardless of the developments in satellite communications services, such fibre optic links are seen to have a valuable role to play in inter-continental information flow. Furthermore, there are plans in Britain, Japan and France for experimental local fibre optic networks which will provide house to house connections on a fibre optic communications main. Just as we have come to expect gas, water, electricity and the telephone to be brought to our homes by pipe and copper cable, perhaps the next generation will have the same expectation for information by optical fibre.

Recent changes in the legislation governing telecommunications in the United Kingdom have paved the way for the establishment of new telephone networks to compete with British Telecom for the lucrative business traffic. One of the most ambitious projects aiming to try and capture a share of this market is Project Mercury. This project intends to lay a network of optical fibre cables alongside the railway tracks which link our major cities. Initially the project aims to connect centres such as London, Birmingham, Leeds, Manchester and Bristol and to provide a link for computer data as well as conversation. The planned expansion of the system will include facilities for teleconferencing and text transmission. It is this latter function which challenges the printed word.

The new information technology is resulting in a fundamental change in the methods of newspaper production. The modern newspaper production system does not rely on the Gutenberg technique but embodies microelectronic technology up to the moment of actually printing the paper. Reporters enter their copy at computer keyboards, editors pick their stories from electronic stores and read them on VDU terminals. Composing may also be done by electronic photo-typesetting techniques. The newspaper itself, though produced more efficiently as a result of the application of modern technology, is still printed on paper and distributed to the public by physical means. But for how much longer?

The term electronic newspaper is used to cover a range of products resulting from the application of information technology with the common intention of providing an up-to-the-minute news service to the public. The simplest relies on a facsimile system for the transmission of script, print or diagrams to permit either single or multiple copy reproduction at a distance from the place of origin. This is not a particularly new technique but one which has gained both speed and efficiency through the adoption of new technology. The British Intelpost system offers a high-speed facsimile transmission service as a public utility between London and other towns and cities in the United Kingdom, America, Canada and Holland. The Soviet daily newspaper *Pravda* uses a fast facsimile system with satellite links to enable simultaneous printing in cities thousands of miles apart.

Perhaps the most exciting new product to emerge from the application of information technology to publishing is videotex. This is quickly becoming an established force as a news and information source and consequently poses a considerable threat to the continued existence of conventional print-based systems. Videotex is the generic name for computerised information systems. These are currently of two types. Teletext is a broadcast system capable of being received by appropriately tuned television sets equipped with a special signal decoder. Viewdata, the other form of videotex, is also picked up on a television receiver but, in addition to requiring a special adapter, must also be connected to the public telephone system which is used to carry the transmissions. Both systems are of British design and are available here as Ceefax (BBC) and Oracle (IBA) teletext transmissions and as the British Telecom Prestel viewdata service. Britain is not alone in trying to exploit the potential of this application of information technology. Videotex is being implemented internationally with such systems as Teletel (France), Teledon and Vista (Canada), Captains (Japan), Datavision (Sweden) and Bildschirmtext (West Germany) on trial or already in operation.

Because the information stored on such systems is regularly updated they are heralded as the first true electronic newspapers. Teletext has been designed with this role in mind and its advocates extol its ability to provide instant information on current events so that users can often have the latest news headlines but also long before they appear on normal radio or television broadcasts. The system is convenient in that it can provide the information when the user wishes it. It is versatile in that the user has the option of having news flashes displayed on his television screen as he watches a normal broadcast programme. Viewdata providers do not all claim to be able to give such a rapid news service as the teletext systems, but because of the scope and design of the system can provide a more comprehensive and detailed database facility.

The information provided by teletext is stored under a system of numbered pages which are accessed by the user entering the appropriate three digit code number on a hand-held keypad. The pages are transmitted cyclically using the extra scan lines of the broadcast services. It takes about twenty-five seconds to cycle a hundred teletext pages. When a page is requested the user has to wait until the next time the page is transmitted in the cycle before it will appear on the screen. Thus, to keep the delay time down to acceptable levels, not all the pages in the system are placed in the cycle at one time. Some pages are available only at certain times while others, once requested and displayed, call up a series of sub-pages automatically.

Each page can have up to twenty-four rows of forty standard-sized characters. Upper and lower case letters, numerals, punctuation and graphical symbols are all available in a variety of colours and large-size characters and drawings can also be produced. Although there is a wide variation possible in the style of page presentation there are system features which are held constant. These are the page number, cycling page number, date and time which are displayed along with the name of the service at the top of each page.

Compilation of teletext pages is handled by a team of reporters, designers and editors in an editing office equipped with electronic data collection and processing equipment. Staff enter and edit copy to a computer store using a keyboard and VDU. Information is drawn from the computer to update obsolete pages and construct new ones which are then channelled from the editing computer to the broadcast one from where they are threaded into the transmission sequence.

At present a teletext receiver costs about a hundred pounds more than a standard television or about three pounds per month more to rent. As sales increase these costs are likely to decrease and it is forecast that by 1983 many TV sets will have the teletext decoder and keyboard as standard features. The Autumn 1981 total of 180,000 sets in use is being added to at the rate of some 10,000 per month. It is predicted that by the end of 1982 a million sets will be in use and that by 1985 this number will have increased to three million.

Prestel, the United Kingdom viewdata service, is based on a system of pages not unlike the teletext services but they are much more numerous (some 190,000 at present) and are arranged as a system of indexed trees. Prestel is accessed by dialing the viewdata computer via the telephone connection. As in the teletext systems, pages of information are requested by entering their digital code on a keypad. If the number of the page containing the required information is unknown then the user may work with the in-built index system to branch down the information tree until the desired end page is reached. The information held on this system is entered by a number of information providers (IPs) many of whom are umbrella organisations for other providers and handle the service on their behalf. Much of the information currently on Prestel is free but there are also pages for which charges of up to 25 pence per page are levied. In addition to any such charge the Prestel user has to pay a computer connect charge and a telephone charge.

About 10,000 Prestel sets are currently in use with 80 per cent of these in business locations. The advantages to business of having access to many thousands of pages of regularly updated market statistics and financial analyses seems clear. With Prestel sets costing about four hundred pounds more than standard receivers and with the add-on costs incurred with usage, the home market has been hard to penetrate. However, as cheap adaptors become available to convert existing TV sets to receive Prestel then this situation may improve.

While the challenge posed by teletext would appear to be largely in the area of instant news provision, that of viewdata is much more comprehensive. The French have already decided to cease the printing of telephone directories and to replace them with a viewdata system. Many organisations are seeing the advantage of presenting their customer information through viewdata. Others are recruiting staff via this system. The economics of providing rail and air timetable information on Prestel are already becoming evident. There seems little doubt that the further extension of this system will present a major challenge to many areas of traditional publishing.

Outside the public system a number of large organisations have installed their own private viewdata systems to provide a paperless information base. Others have formed closed user groups within the Prestel system to share particular collections of pages. Both these applications of information technology dispense with the need for many in-house company publications.

Education has also recognised the value of viewdata. Apart from pages describing courses and giving details of course vacancies, the current database also provides a number of programmed learning sequences. The provision of learning materials in this way is thought to have great potential for distance learning. One noteworthy example of a Prestel learning sequence has as its subject the basic principles of offset lithography and has been produced by the Print Industry Research Association!

The Prestel system not only provides information to the user but also has the facility to collect information from him via a system of response frames. This has been used by a number of organisations, particularly mail order companies, to permit purchasers to place their orders directly rather than by over-the-counter transactions or conventional mailing. The Prestel system recognises users by a unique sign-on code and receiver identification number. Simply keying in a valid credit card number against a selection secures the purchase. Not only can the user return information to the system but he may also communicate with other users, providing of course that the sender knows the sign-on code of the intended recipient of his message. For users with only a numeric keypad, messages must be coded onto preformatted frames, but subscribers with a full alphanumeric keyboard have the ability to construct a complete frame of information. This electronic message service (Mailbox) is currently only available in the London area but with the introduction of the Prestel Advanced Network Design Architecture (PANDA), the Mailbox facility will be extended to the British Telecom computers in the other regions.

The next major step in the development of a public electronic information utility will be when the PANDA networks will have access to databases and computer systems other than those of British Telecom. This will be possible via a system of gateways which will control channels to a variety of computer and information handling systems. This will also open up the real possibility of genuine electronic funds transfer. Rather than, as at present, keying in a credit card number to make a purchase via the response frame of the viewdata system, there will be automatic access to the purchaser's bank account with a credit check before confirming the purchase and debiting the account. The West Germans already have a system of this type on operational trials. This electronic funds transfer (EFT) involving the implementation of financial transactions by electronic means is a considerable step forwards from the now quite common point of sale (POS) systems which merely record and report cash or credit transactions by means of electronics.

There has been some progress towards EFT in this country. Following experiments involving Barclays Bank and a number of garages in the Norwich area, the chief executives of the five London clearing banks have now agreed on the outlines of a cashless transaction system. The system will be based on point of sale terminals in sales areas which will communicate directly with banking installations via telecommunication links. Purchasers of goods would initiate payment at these terminals by means of a plastic credit card. The debiting of a customer account and the crediting of the retailers account would then follow automatically. The plan is to set up a number of distributed networks of regional computers to handle the transactions but to link these by means of the British Telecom system. This would have the advantage of allowing individual banks to progress with the adoption of the technology at their own pace yet retaining the compatibility necessary for future expansion. A potential outcome of the successful implementation of such a system could be the elimination of cheques and paper money from our society and, along with that, the paperwork presently necessary to check and verify purchases and sales for cash or paper credit. If such electronic funds transfers can be coupled to computerised ordering then, not only is a great deal of paperwork eliminated but also there is created a better management information service on such matters as sales patterns, customer payments, supply efficiency and stock demand.

With the adoption of technology which holds copy in digital, machine-readable form on magnetic media, there is the possibility for rapid and widespread transmission. When coupled to suitably swift printing techniques such as laser printing, with a capability of up to 20,000 lines per minute, the possibility of on-demand publication is indeed a real one. With the market for some publications such as reference books, directories, timetables and annual guides being short lived, volume production may be inappropriate and even unnecessary. There is a case, not only for on-demand publication but also, with particular regard to the publication of transient information, the elimination of the printed page and its replacement by the electronic one. The establishment of teletext and viewdata services are the first steps which have already been taken in this direction. However, such services must not be seen as mere replacements of existing services but as important ways of extending communications and opening up new areas for development.

One of the barriers to the further exchange of information between computers is the variety of ways in which data are stored. Groups of big computer users such as airlines and banks already arrange data in the same format for convenience of internal transmission but so far the international standards organisations have failed to agree on a single common system for organising computer-held data. There have been standards established for the way in which information flows between computers and for the connection of circuits and wiring but not for data. The British Standards Institution have suggested a new language (DIAL) with its own syntax and grammar to enable data records to pass freely from machine to machine ensuring compatibility of content. If this can be achieved then it opens up the very real possibility for the types of information exchange which could eliminate much of the routine paperwork of business and commerce and see the establishment of a universal electronic mail system.

While few would agree that the end of the printed word is yet in sight, many see the future of traditional publishing as lying in the area of multi-media packages. This is thought to have a particular market appeal and a number of projects are being planned. Among these is a joint venture by Thorn/EMI and Mitchell Beazley to combine the well proven low technology of the book with the largely untested and high technology of the video disc in a range of home instruction packages.

There are a number of other developments in the field of electronic publishing which are aimed directly at the ultimate replacement of printed text. One such development has stemmed from the high production costs incurred by a number of learned journals as a result of their limited number of subscribers. If the function of such journals is seen to be in the rapid dissemination of current research then, as such, they seem appropriate subjects for the application of the new information technology. Different examples of such applications can be suggested. Text prepared by contributors on word processors or text editors and supplied to the print room on floppy discs need only be arranged in a suitable format prior to the production of copy. A current experiment by the British Library carries this idea a stage further and seems set to pioneer the first true electronic journal. The new journal, *Computer Human Factors*, will be compiled entirely electronically. The normal processes of writing, refereeing, editing and publishing will be carried out with the aid of inter-communicating computers. Contributors, referees and editors will all provide their inputs and channel their communications through computers in varying locations using devices such as optical character readers and word processors.

Although the electronic journal as yet seems unlikely to completely replace the accepted paper periodical and there is little immediate prospect of trying to convert academics from paper readers to VDU scanners, one other aspect of electronic publishing is worthy of note. That is the growing number of on-line information databases which are being increasingly used to locate published items of particular interest.

The British Library's Automated Information Service (BLAISE) provides one such information retrieval service from a number of databases. BLAISE provides access to four types of database. These consist of bibliographic databases which hold citations to published literature on various topics, MARC databases holding information on published works in a cataloguing format, databases with information on specific research projects, products and materials and dictionary databases designed to provide assistance for on-line searching. The BLAISE information is held on a computer located in Harlow, Essex but can be accessed via a standard telephone linked up with a suitable terminal. The system is also accessible internationally at a low cost through the Euronet and IPSS communications channels. The information which results from a search is displayed on the computer terminal with an on-demand printing option. With the likely expansion of such services one must ponder on the wisdom of the continued publication of full text articles in many specialist journals when there is the possibility of providing copy to interested enquirers on demand.

There is at least one publisher who is offering full text electronic publications capable of being read at a distance and printed on-demand. This system of telepublishing has been established in Britain by Butterworth (Telepublishing Ltd). Their Lexis legal information retrieval system which is on offer to both lawyers and accountants, allows telephone line access to millions of pages of case law, statutes and other legal reference material held on computer. Full-text searching is permitted on a user's selection of key words and the text surrounding the key words is displayed for inspection on a VDU terminal. The user can move rapidly through the textual material choosing to retain only those pieces of particular relevance. There is the facility for full-text retrieval and for obtaining a hard copy output of selected text. There is little doubt that this is just the forerunner of many such systems soon to appear and to be offered not just to the professions but also to industry, business and commerce. There would also seem to be a particular application for such facilities in the public information services, in libraries, schools, colleges and the homes.

Just as the electronic typewriter has gradually replaced the manual one so the word processor will eventually replace it. The typical word processor has a standard QWERTY keyboard but outputs the entered characters onto a VDU screen rather than to paper. This allows for rapid checking, editing, correction and reformatting to be done prior to the production of a paper copy. This ability in itself offers a considerable advantage over conventional systems, the speed of output of which is limited by the time taken to correct mistakes. However, a greater advantage is achieved through the processor's ability to store information on its computer discs. This not only eliminates the need for retyping after further edits but confers the ability to produce personalised standard letters by automatic means and creates an electronic filing system. The word processor operators are much more than copy typists. They are filing clerks, stock controllers and communications mediators. Their role is crucial to the efficient functioning of the electronic office.

The advent of the electronic office will not only replace the current paper based office technology with electronic data processing methods but will also open up and streamline the communications channels of an organisation and so provide the potential for a more efficient management information service. An integrated system of word processor, local memory storage facility and communication gateways to databanks would seem to be capable of efficient operation without much recourse to the use of conventional printed material. Such a system would facilitate the transfer of electronic mail and provide a system to file and retrieve information with ease. The adoption of such an integrated electronic office system would seem to be capable of confering several advantages to an organisation. Even now, electronic communication is priced competitively with conventional hard copy mailing and carries the added benefit of speed. Furthermore it eliminates the need for operator re-entry and organisation of data and consequently has no need for either the high cost of labour or the accompanying resource support.

As part of the plan to provide practical examples of the application of information technology in the Government's 1982 Information Technology Year, a section of the Cabinet Office will become one of the country's first paperless offices. The new Information Technology Section of Downing Street will be the first to benefit from the local area network (LAN) link which will connect a number of word processors, computers and other electronic devices. The intention is that terminals will be placed on executives' desks and they will be encouraged to draft and edit their own reports, call up information from the computer databanks and communicate with each other via an electronic mailing system.

The applications of the technology seem endless. We now have the 'teleputer' which is a combined personal computer TV, video recorder, typewriter and database. As the market for such devices is created and encouraged to grow, so the price will fall and, just as we have other communications devices in our homes so we will have our own teleputer. As was the case with pocket calculators, the market will be created. The impact of the widespread possession of such powerful and versatile technology will be great and is likely to occur at a much greater speed and be of much greater proportions than that which accompanied the communication revolutions of yesteryear. The wired society of the not-too-distant future will have the electronic services which are now emerging and which undoubtedly will be supplemented with others even more powerful. People will listen to and read electronic books and newspapers, consult electronic files and databases and communicate one with another via a future generation of teleputers. While there is little doubt that the domination of the printed page as the vehicle for communication of text will be challenged and probably usurped, there seems little likelihood of its complete demise: print gives too much pleasure to so many people for this to be allowed to occur.

2. THE IMPORTANCE OF INFORMATION IN INFORMATION TECHNOLOGY

JON M. MASLIN
P.I.R.A., Leatherhead, UK

So far, in all the discussion of information technology, the emphasis has been on the technology, and less emphasis has been on the information and its creation and use. This is not surprising. The far-reaching innovations in technology make possible changes in the nature of handling information and the need to re-examine the treatment and the management of information follows. Despite the sophistication of much of the technology the problems associated with making it do something that is wanted by people is in many ways more complex. The efforts required to achieve this are going to involve everybody and take a much longer time to develop than the technology. It will be the technology-led revolution for a number of years yet, but there is growing feedback from users and providers of information which will modify the way the technology is applied.

The two majors trends that have already become apparent will continue. The relationship between cost and performance will continue to improve in the user's favour, and we can expect improvements in the tools available for creating, transmitting, storing, manipulating, retrieving and displaying various permutations of information such as data, text, graphics and speech. This represents the much referred to convergence of computer, telecommunications and information processing technologies. Evidence of this is seen in the breakdown of traditional barriers between the computer industry, telecommunications industry, office products industry, the television industry (equipment and programmes), publishing industry, printing industry and others.

Great store is being set by the 'information society' or

'post-industrial society' in which much greater emphasis is placed on handling and processing information, controlling manufacturing, and controlling and administering resources (or for leisure). Implied in this is the emergence of different ways of conducting economic and social exchanges and so different ways and patterns of work. National governments are putting considerable effort into developing the technology to do this as a means of continuing growth, or breaking out of a stagnant economy and society, or at least maintaining control of a country's trade and identity. One of the most salutary examples of this was the now well-known French 'Nora report', *L'information de la Société*. The French government has adopted a strong, centrally directed role. The West German government has also implemented successive programmes to develop the computing industry throughout the 1970s. The UK government has similarly recognised that information technology is a vital part of the future and the Minister of Information Technology has allocated £100 million towards the development of the UK microelectronic components industry, £80 million over four years to promote awareness and use of information technology, and has announced that 1982 will be Information Technology Year. On a European scale the EEC is intent on fostering information technology and the interchange of information.

The USA has not formalised its approach as much as these countries, but it dominates the information technology sphere with the leading companies, a vast home market and large public procurement programme.

The Japanese in particular have decided on the importance of information technology to their future and back in the early 1960s started developing plans. Japan provides probably the best example of an information society. Jean-Jacques Servan-Schrieber, in *The World Challenge*, describes how Japan's post-second world war growth has been based on exploiting its only major natural resource, human intelligence, which he claims becomes more productive year by year. Despite, or because of, the cataclysmic change in Japanese society, he argues, there is a collective quest for knowledge and Japanese society is permeated with all forms of information gathering. This has led from the collection of information around the world on technologies and market needs to putting it together and interpreting it, to the decision to develop a major information technology activity. In the process high levels of automation have led to an increase in average personal income from $20 in 1945 to $12,000 at the beginning of 1981.

An essential characteristic of human society is the ability to organise all forms of knowledge and as a result improve understanding. What information technology does is to enable many more people to do this with much greater power. It is a continuation of what has gone on before but represents one of the few major leaps forward – requiring changes in the way that information is conceived of as a resource. It will need very different ways of creating it, structuring it, retrieving it and especially valuing it. Within the context of the broader issues of the information society, as described by Daniel Bell, it may be useful to look at some of the more practical issues that are already in evidence in the communications infrastructure that he considers will be subject to the most profound and far-reaching effects (the other infrastructures are transport and energy).

We already have a number of the embryo stages of publishing information which demonstrate the application of new information technology. Some have been designed from their inception for specialised groups of users, some for mass use. On-line bibliographic databases, largely for scientific and technical topics, were first put on-line for public access in the later 1960s. Since then their scope has increased to include applied sciences and technologies and softer sciences such as sociology, business, marketing, news, advertising, etc. The Eusidic Database Guide 1981 shows that the number of databases available to users in Europe, including those available from the USA, rose from 335 in 1975 to 654 in 1980, and the number of databanks rose from 51 to 75.

The bibliographic services demonstrate the interaction between technology pull and demand push. They provide instant reference to the information required, but usually stop there. They do not provide answers, so ways are being developed to provide on-line ordering and delivery of documents, which have not been without their problems.

Much of this investment in technology is aimed at commercial and business information transfer, resulting in what is being called 'the electronic office'. But it will also be applied to all types of publishing – for business, education, research and domestic use. It is this area of creating and using information services that is crucial to the uptake of the new technologies and yet, so far, insufficient attention has been paid to it. The adoption of new technology implies a great change in the way people think of information, the ways they use it and the methods in which they pay for it. This change is going to take place at a much slower rate than that at which technologies can develop.

From the users' points of view it means that they will have a much greater choice of media, and that they will be able to choose what is most appropriate to their needs. This will be determined by the medium that is most appropriate or attractive and by the type of information and the use to which it is to be put. For example, a statistical databank may be provided with economic modelling facilities, a solicitor may interrogate a full-text database of case law, a student may use video disc with random access interactive player or a travel agent may use a viewdata service for travel information. The days when print and, occasionally, microforms, were the only media have gone, and the number of alternatives will increase. Publishers will have to consider more carefully the uses to which their information is likely to be put by different users in order to present and package it appropriately. Basically the same information may be provided by different media for different users but it will have to be treated accordingly. Statistical information may be provided in relatively unprocessed form for users who want great detail and have the facilities and resources to manipulate it, while it may also be presented in a form with much more editorial selection on a viewdata service for users without great demands or processing abilities.

Users will also have to invest in equipment in order to receive and process the information. The complexity will depend on requirements but is likely to take the form of a telecommunications linkage, a display screen, microprocessor intelligence, memory and some form of hard copy output device. Again, there is choice. The need for equipment introduces an important new dimension because users will only buy it if there is information available that they want. Conversely, until receiving equipment is widely installed it is not economic to provide information, and until large markets can be created the equipment remains expensive. This is an obstacle that will require great skill in overcoming by creating services which will stimulate the market and create a momentum on which a large number of other services can be justified on their marginal costs. Compatibility will be achieved by telecommunications interfaces and extensive international standardisation, as for example is progressing at present in the viewdata and teletext areas.

Users will also be paying for the information in a different way, with probably greater emphasis on buying small parcels of information as required and with the immediate user paying a greater proportion of the direct costs. Changes in the financing of information are highly significant. No matter how information is stored and retrieved it is expensive; electronic publishing tends to accentuate the value of pieces of information rather than spreading costs throughout an infrastructure, including book distributors and libraries. Users will also have to bear more of the direct costs of distribution because they will often be paying telecommunications charges and possibly computer time.

Users will also be obtaining information for further processing or manipulation either for profit or pleasure. The natures of the media available will mean that in many cases what is required will have to be defined with greater accuracy than previously and that conscious decisions will have to be made to obtain it and pay for it. Correspondingly, publishers will be making great efforts to make information available. It is likely that in many cases it will not be as difficult to obtain information as to filter out what is not required.

Development is likely to be gradual, with users and providers progressing step by step as the publishers demonstrate what they can do, learning from users and stimulating them to develop their requirements. The need for information exists, but it has to be conceptualised and put in a marketable form – a job which has always been the function of a publisher.

One of the great problems is not in establishing what information technology is, but what information services people need. At present there are gaps between the developers of the technology, the developers of systems and information services and users. To a large extent this is due to a lack of understanding by users about what is possible. This could lead to systems being forced on them which are unsuitable and which are rejected. It is obviously difficult in such a situation to find out what people want, because they cannot answer. If you are unaware that it is possible for planes to fly, and do not care much anyway, it is not much use being asked for opinions on the merits of piston or jet engines.

We are now at the stage of establishing a technical base which should demonstrate some of the things that are possible so that users can develop an understanding and make demands on the providers of technology. The result of this is likely to be a mixture of different systems which intertwine. The ideal way, of course, is to start with the definition of a system to meet the demands of the user and then develop the system, but given the iterative nature of development this will be difficult to achieve. The mixture of different systems is likely to lead to an unnecessary complexity as paths of expediency and individual cost reduction are followed, whereas widespread acceptance of information technology will need to rely on simplicity in use (although considerable sophistication may be needed to achieve this).

It will be important for the systems to respond to the users as individual people. This means that it will be necessary for the user to access public databases, databanks and computers, private databases, databanks and computers, probably with some means of interlinking and to connect these with personal or corporate processing and storage facilities. It implies that users will be able to get different types of information – text in page form or more suitable formats, document facsimiles, graphics, maps, statistics – manipulate them and store them for their own applications, make copyright payments, etc. They will then be able, in turn, to publish or distribute the results either publicly or privately for a specific individual or group. It also implies that users will have different terminals and processing facilities according to their needs. Some may have no more than an upgraded television set which will access a viewdata network (and gateways), others will have a personal computer, and others intelligent terminals connected to powerful processing and storage systems, and printing devices. This gives the user the choice of what level of sophistication he requires for level of display, interaction and processing.

Different types of databases and searching methods may cause problems, particularly for the non-expert user. The change in techniques and habits required could prove a handicap unless considerate guidance is given to avoid failure or a sense of frustration. It is very much the responsibility of the service and information provider to make the systems compliant to the user.

For many people viewdata has provided the first experience of electronic publishing and has shown that there are alternative methods of presenting information. Like most innovations the methods used in previous technologies have been carried through, just as early cars owed much to horse carriages, and early cast iron structures (such as Ironbridge) to wooden structures.

The change implicit in viewdata, though, is greater because it is two-sided: the information provider has to develop the way he uses the medium and the user has to learn how the medium is being used and become accustomed to it. Both have to develop in harmony. Compared with innovations of similar magnitude in media, such as cinema and television, viewdata is still in its infancy and there is much to learn about the types of information, the way it is put together and the design of an overall service and individual pages. Television, for instance, thirty-five years after widespread broadcasting started, is still developing methods of presentation, content and viewing patterns. One factor alone is likely to make the innovation pattern for viewdata more fraught and that is that it follows a very well established and developed medium – the printed word. Television has had to follow the cinema, which was not so well established as the book, and the cinema followed the theatre and music hall, neither of which had the pervasiveness of television. Neither has demanded such a change by the user from previously accepted methods.

The treatment of information for viewdata is different to that of any other medium, and most information providers have found that existing information has to be rewritten or reformatted. The main reasons for this are the dividing of information into small pieces to fit the page structure, the relationship between the pages and the limited space available on frames.

Other reasons are the availability of colour, the limited variety of type sizes and the nature of the reading process from the screen by the user. As is now well known, the basic structure of the viewdata database is hierarchical with each page having ten filials so that an information provider has a pyramidal structure of holes which he can fill with information. Within this structure up to ten routes from a page can go to any frame. The internal structure of the information need not, and probably will not, follow the hierarchical database structure. The skill of the information provider is in creating a structure which approximates to the likely needs of the majority of users. The information provider has to put himself in the mind of the user so that he can create the routes to the information needed. There are often likely to be several routes to a piece of information to answer different needs or way of thinking about the information. Particularly with a system like Prestel, where the information provider pays to rent each frame and often obtains some of his revenue from page charges, there is an incentive to open up as many routes as possible so that information page use is maximised. However, routeing pages impose a heavy load on the information content. It is not unlikely for a quarter or a third of the number of frames to be routeing pages. These do not contribute to the revenue but just act as signposts and, to varying degrees, get in the user's way.

The routeing structure has provided one of the gripes from a sector of opinion that argues that it is tedious to work through a sequence of routeing frames, making successive choices to the information that is there (or not there). While the disadvantage of this can be overestimated, it does represent an important example of the friction that can develop between the expectations and needs of a user and the aims of a system provider to give a simple system and keep costs as low as possible. It also represents an example of the change in working method that is required. Even if it requires ten routeing pages to obtain the information, it is likely to be as quick as finding a printed source and then finding the information within it. One of the questions to be answered is how far the user should be expected to change his method of working. There is a danger that we will assume that the old established ways are intrinsically right, when they are only adaptations to the available technology.

Conversely, if the technology is forced on users there will be a rejection of it, so there must be a continuing process of adaptation. One of the areas where there needs to be improvement is in indexing, and efforts are being made to automate the production of indexes and to provide keyword searching. The problem with keywords, of course, is that keywords can have very different meanings according to user and context.

The limited capacity of a viewdata frame can be used to positive advantage as wording has to be concise and ideas and facts broken down into small elements and presented simply. Combined with the ability to present a logical structure where information can be accessed in the order relevant to the user, it can provide a powerful and effective way of presenting information if the overall concept is carefully established and attention paid to routeing and writing. Because of this combination there is an important change in the creation of information – one person may be involved much more personally with the whole product from conception, to design, writing and inputting. There may need to be much less compartmentalisation than has previously been experienced.

Another important difference is that viewdata is a much more interactive medium, and more positive methods have to be employed to lead users on to access the next frame and spend more money. A number of means are available to do this. Good indexing is important, but not the complete answer, and imaginative cross-linking is important but difficult to arrange between separate information providers.

With the trend towards marketing for specific sectors, it is possible to see integrated services developing. This would provide information on several levels. As an example, one level would include regularly updated information which would encourage users to develop a habit of referring to a database to get the latest bulletin. From this would be prompts to related detailed information as a way of promoting access. Information that is likely to be accessed through a specific need would have a route to a current news item, again to stimulate use. In some cases the detailed information will not be suitable for viewdata treatment, and the user will be led through a gateway to a third party database where there is a larger store of information and more powerful retrieval software.

Prestel provides a salutary lesson to anyone considering publishing. Why are the numbers of terminals installed so far below the estimates of a couple of years ago? The glib answer is that the terminals are too expensive, but this is not the whole truth. A major reason for this is insufficient demand for the type of information being provided, or its corollary, the demand not being satisfied by the information provided. If there were a demand for the information, viewdata terminals would be bought in just the way that video recorders are now being bought. The price of a video recorder is greater than that of the addition of viewdata to a television set, and yet the rate of growth of video recorders is equal to or greater than that of colour television.

Despite the realisation right from the beginning of the importance of the tripartite nature of viewdata – an alliance of system operator, the television industry and the information provider – the difficulties of creating information that users want in the form most appropriate to the medium, and of creating a demand in the minds of users, was greatly underestimated.

As many information providers have found, and are continuing to find to their cost, it is a long and expensive business to develop the right information, structure and design it, and identify the right market, develop it and create the right price structure.

Information has to be organised in a different way, written in a different way and displayed in a different way. New skills have to be defined and learned; and multi-skilled practitioners produced.

Looking at some of the services that may have a significant role, newspapers could provide a new conjunction of entertainment and information services. Rather than thinking of themselves as producers of printed newspapers for which they collect information, there will be a change of emphasis so that information is stored and handled to enable it to be distributed and sold in a variety of ways, including the production of a printed newspaper. Some of the elements needed to make this change are already being implemented, but the combination of the new approaches will be significant.

Newspapers are essentially ephemeral, but their contents

are often of lasting value, so considerable effort is being made to provide services which enable the valuable stores of information built up by newspaper publishers to be made available more readily, both as an aid to help its writers obtain the information they need and to generate another form of revenue. Most progress has been made in America, but the *Financial Times* in London has an on-line database of company information from its newspaper. In the USA the *New York Times* Information Bank has probably the best established database service. Starting with a morgue operation in 1973–4 for the *New York Times* itself, the Information Bank has developed a general interest database with some 125 other publications being abstracted. The marketing strategy is to get people used to using a general database, then sell them and develop a specialised database, for example one specialising in the Middle East was established in March 1981.

Most of the activity so far has been aimed at the business and professional user where there is the demand for information monitoring and retrospective searching. For the domestic consumer less progress has been made, but the argument for seeing the newspaper, particularly the local newspaper, as an information gathering centre is equally valid. There are large gaps in the way that local information is provided which could be filled by a local centre which can be interrogated easily. These gaps result from the information not being available where it is required, but located in a variety of places, or not being assembled in a useful way. For example, information published in a newspaper two or three weeks ago is rarely available without it being cut out and stored (unusual in most homes).

To a large extent the factors which have prevented these gaps being filled have been the difficulties of storing and distributing the information economically. The use of computer databases, telecommunications and cheap terminals open up the prospect of being able to collect and store a range of information to which users can refer and select the information they need. Again, the user would be offered the opportunity of choice. The choice would be at least as wide as the total content of the local newspaper, but it would be searchable, for example it would be possible to find out the decision of a council on a planning application, when a local event was planned to happen, and local suppliers of goods or services.

This type of facility could be structured in a number of ways. It could be the responsibility of one body, such as the local newspaper (which would probably be undesirable because it would give control to one organisation); or it could be organised by an 'umbrella' organisation, facilities being provided into which a number of independent providers of information could place their information; or it could constitute a number of independent databases, which would offer freedom of choice, but also possible duplication and more difficulty for the user. The second choice probably offers the greatest potential because it would provide for the greatest range of information and processing facilities, and a unified (and therefore easy to use) structure. It could also offer the most democratic and human approach, and would enable narrow market sectors to be addressed and people with minority views and interests to be catered for by enabling them to have a means of distributing their views or information at a more economical cost than existing printing distribution channels. The role of the 'umbrella' would be to provide the local computer, and overall database structure and maintenance so that users and contributors could find their way about it.

How practical is this? The answer is that it is not yet really practical, but that already it is happening in part in the USA and Europe. In the UK a number of newspapers have developed their own local electronic newspapers on Prestel, initially as a defensive move to protect their markets against a threatening newcomer. None of them has made any money and they have all moved on from their original concepts of how to use the new media to providing a range of information services in addition to news including local information, national feature material, and viewdata consultancy services. However, they have all suffered from the slow uptake of viewdata terminals by the domestic market. Despite policies in some cases of not making a page charge, the value of the information has not been sufficient in users' minds to make them want to buy a viewdata set.

Encyclopaedia publishing is an interesting area which demonstrates some of the possibilities in electronic publishing. It needs to be considered in the context of a number of trends: trends towards packaging into convenient parcels, characteristic of a great many articles and services; and the opportunities offered by computerised handling of information in many modes. The Arete Publishing Company in the USA, a subsidiary of the Dutch publishers VNU, is a good example.

The 'Academic American Encyclopaedia', which consists of nine million words in 32,000 entries, is conceived of as a database which can be updated as required and from which spin-off products can be created. Arete claims that 50 to 85 per cent of the material for these spin-offs will come, virtually without further modification, from the encyclopaedia database. Because all the articles are coded according to their discipline and sub-discipline and several other classification schemes, it should be a simple task to extract all the material relating to a particular topic. This enables Arete to create new products, such as a book on a specific subject, with a minimum of editorial and production operations. If necessary, extra information or editorial content can be added. The output can then be sent directly for typesetting. The product need not, of course, take the form of a book. Text could be sent directly to a purchasing organisation over telecommunication links for printing out on a high quality intelligent copier such as a Xerox 9700, so providing course material for a lesson, for example. With the addition of sound and filmed material it can also form the basis for productions on video disc.

For this to be possible, the internal structure of the database has to be organised carefully, with articles following a strict and pre-imposed pattern. These patterns are different from those usually found in newspaper articles, for example, where the material is arranged in descending order of importance. The disadvantage of imposing this rigidity on the contents is a loss of freshness and a certain sameness from article to article, but this may be regarded as a minor price to pay because of the way people commonly consult encyclopaedias.

The approach also has an effect on the way the encyclopaedia is kept up-to-date. Changes can be incorporated much more easily, or separate updating services can be provided through viewdata services.

While much information technology uses telecommunications or cable for distribution, there are other media which are important or are likely to become important: magnetic media such as floppy discs, tape cassettes, video tape; and video and optical discs. All have become established means of distribution except, so far, the video and optical discs, and the application can be expected to increase. Indeed the growth rate of the market for video tape recorders has, in many countries, exceeded the growth of colour television, with applications ranging from recording television programmes to cinema films and education and training material. Video tape is basically serial in access, but considerable progress is being made in enabling them to be used non-serially by fast searching methods.

Greater potential lies with video discs which, in their interactive forms, have moderately fast random access capabilities so that when a player is connected to a microprocessor or computer, the user can interact with the programme provided. The programmes can be created with a combination of moving colour pictures, diagrams, still pictures, text and sound on two sound tracks, giving either stereophonic sound or two different sound tracks. Programs to control access to the disc can be either held on the disc or supplied separately, so enabling material on a disc to be used for many different applications, thus spreading the high cost of origination.

A prime area of application for interactive video discs is in education training, do-it-yourself instruction, etc., where a wide range of material can be combined and the user can interact with it personally.

While arguments exist for the effectiveness and value of computer-assisted learning, self-instruction, programmed learning, etc., what is certain is that success depends on first-class material and programming that is carefully tuned to the user's requirements. This requires skill in understanding the learning process, writing, programming and visual presentation, and also management skill in forming an effective team of creators. Without this the technology is worthless.

It is probably worth saying that none of these media should be considered in isolation. Video discs may be used in conjunction with personal computers, with viewdata, teletext (linking telesoftware and video discs perhaps). It is also worth pointing out that they will not replace conventional methods, but that they will be used in conjunction with them.

An integral part of publishing is the creation of information, and this too will be affected. Already authors are writing directly on terminals connected to computers, word processors or personal computers. In the scientific field, electronic journals and editorial processing centres are being experimented with. The importance of this trend is that it gives the author much greater control over the final appearance of his work, and in some cases the opportunity for interaction with his readers. A question that arises is how will the new media affect the way in which people present information? With much improved facilities for non-serial access, and a different mode of presentation, will conventional methods persist? There will be strong establishment pressures from some quarters for them to do so. In the case of viewdata, not only does much information have to be restructured, but the creation of individual pages is much more effective when it is done directly on a screen. The terminal will assume a central role in information handling, both for retrieval and creation. There are increasing signs that people who, a few years ago, would not consider using a typewriter, are now prepared to use a terminal.

An area which is of fundamental importance to the implementation of information technology is the relationship between content and visual presentation – the 'language element' as it has been called. So far it has been neglected in the concern about technological progress. The language element is closely related to the particular technology being used, yet it remains constant in terms of its function. The technology affects the way information is arranged spatially (ranging from linearly to non-linearly), and the symbols used (such as words, pictures and schematic images). Information technology will have a revolutionary effect on presentation because information can be broken up into discrete units which can be accessed in a non-serial fashion and various types of information can be combined, for example text and diagrams, text and moving pictures, pictures and speeds. The mode of expression has always been modified by the technology available, and the new technology will enable some of the constraints we have come to live with (and like) to be overcome. Styles of writing will also change, using techniques of information mapping, and perhaps veering away from discursive writing to more 'telegraphic' approaches.

One example of this that is already apparent is the news on teletext and viewdata which is treated more like television news than newspaper news.

Another aspect of this which is controlled less by the information provider than by the equipment supplier is the design of equipment and the images, such as typefaces. The information designer can control the layout of the presentation but not the overall appearance.

The importance and complexity of the choices opens up a range of possibilities for which a new set of skills is required, but for which there is no comprehensive training. The skills encompass those used by writer, editor, graphic designer, computer programmer, artist, information scientist, television producer, etc. But there are much wider implications because the whole of society will need educating into new ways of obtaining and creating information. There is a greater need than ever before to study the complete presentation of language and meaning. This is because our systems are fast becoming interactive and are reaching into most walks of life. Increasingly, control of the appearance of information will soon be in the hands of lay people. An example of this is the increasing use of word processing origination for printed matter, where the influence of the typographer and compositor are waning.

Therefore more attention will have to be spent on training people to use, appreciate and create and design information. Education for this purpose must start in the primary school. Precise practices are taught in schools in relation to the organisation of writing that will no longer apply when children grow up; nor do they necessarily apply to other means of production. There should be more concern with teaching fundamental principles or organisation of language which holds goods over long periods and across technological boundaries. This needs a great deal more study in a range of disciplines, many of them basic to the understanding of the human brain.

Intimately linked with the provision of information and the means of distribution are the economics of the new services. There are two aspects to this. On the one side are organisations which want to provide information and are prepared to pay for doing so, e.g. advertisers, and pressure groups. On the other side are organisations which provide information which has to be paid for by the user. Whether the ability to retrieve information will be a sufficient incentive for the home user to ensure regular use of the new services is very much an unknown factor, some people arguing that domestic users, unlike those in the business sector, will not find paying for information in small items attractive. This is where the organisations who are prepared to pay to have information made available come into importance because part of the revenue generated could finance other services, or they could directly sponsor other services. However, there will also be many applications where the cost of the information will be outweighed by its value to the user, plus its timeliness and convenience. There is yet another reason for valuing an electronic medium above alternatives: the ability to interact with the system by sending messages, placing orders, doing calculations and so forth, and retrieving data which can be stored and handled locally.

There is an additional factor which could be crucial to the development of information services, and that is the legal framework of copyright, international data flows and privacy.

The use of information technology offers a number of contradictions: a much greater choice of media, but a commitment to paying for equipment; much greater processing power, but greater skill needed to use the information; greater decentralisation and personal access to information, but possibilities of economic and political control; and the availability of much more information in a useful form or severely restricted information freedom. It is important that the users of information decide where and how the technology will be used.

3. REDESIGNING JOURNAL ARTICLES FOR ON-LINE VIEWING[1]

MAURICE B. LINE
Director General, British Library Lending Division,
Boston Spa, UK

This paper is based not on any special knowledge or research but on a keen interest in the transmission of primary information and an even keener concern with the need to design information systems around human beings rather than the other way round.[2]

The paperless society, or less dramatically the electronic journal, has been prophesied for some years now. Most of the prophets have also been strong advocates, detailing all the disadvantages of print-on-paper and showing how electronic transmission of text will overcome them.[3] It is not always clear what sorts of documents the prophet-advocates are referring to. They sometimes appear to write as if all documents were scientific or technical articles, and certainly the arguments they use apply mainly if not exclusively to such articles, presumably mainly because they are themselves part of the large but limited community that writes and reads them.

Let me then summarise the arguments for electronic storage and transmission, with special reference to scientific and technical articles.

1) Articles once written have at present to be typed, submitted to journals, refereed, often revised, and then printed after a delay of anything between six months and two or three years. In an electronic system, 'typing' would be done on-line by the author (or his typist?); submission, refereeing and editing would all be done on-line by the journal editor(s), referees and author; and the article would then be ready for public access.
2) Articles may contain errors or be superseded by further

research. Some journals publish critical or updating letters, but the original text remains untouched. In an electronic system articles could be revised, presumably under strict control.

3) Many journals are facing financial collapse as costs rise and the market declines with the reduced purchasing power of libraries. To date most journals have survived by means of hidden subsidies (unpaid editors and referees, use of university presses at favourable rates, etc.) or by such measures as splitting into several parts, but this process cannot long continue. With falling electronic storage and transmission costs and greatly increased availability of terminals, the electronic journal will soon be more economic than the printed one.
4) Libraries are running out of space, and in any case modern paper has decay built into it. The permanent and very high density storage offered by an electronic system would solve these problems.

Against these advantages there are several potential disadvantages. Control over input into 'journals' (which may quickly lose their individual identity in an electronic system) will be more difficult but very important if the system is not to be an electronic dustbin. There are serious dangers in the ease with which authors, or anyone else, can change what has been written. The resulting problems would not merely be of concern to the historian of science, since 'correction' of the past and the expunging of records could represent a major political and social abuse.

As for economic viability, a continued growth rate of current journals of 3 or 4 per cent per annum does not suggest a serious crisis, and in any case most high quality journals are under no threat for the foreseeable future: libraries will buy them whatever they cost (provided of course they are within their subject field). What are vulnerable are journals that receive little use, and these are the first to be sacrificed by libraries. However, if they are stored and transmitted only in electronic form every use will be exposed, and so will the whole economics: few hidden subsidies will be possible, and little income can be expected from articles that very few people read. Cheap though storage for on-line access may be, a journal that is not viable in printed form could well be even less viable in electronic form. (One obvious alternative is a synopsis journal with on-demand supply of full text from a centrally stored original).[4]

The technology incidentally still has some way to go,[5] particularly in the satisfactory and economic transmission of illustrations other than line drawings. Since a very high percentage of all articles and of all usage is in biomedicine and related fields, clearly this problem must be solved.

Let us assume, however, that economic and technical barriers will be overcome, and consider the reader of on-line journals. He can of course cheat the system in a sense by having a print-out made at his end, and I suspect that in many if not most cases he will wish to do this, provided the print quality is not too hideous, since he will not want to do all his reading at a terminal. Nor indeed will he easily be able to do so, even if he has a terminal both in his office and in his home – print-on-paper is at least portable and readable anywhere. Another reason why he will want a physical copy is that he is used to reading in this form, though this may not apply to the next generation. Very little research has, to my knowledge, been done on the way people read, and I therefore have to generalise from my own experience and from discussions I have had with colleagues.

What I virtually never do is to read an article from beginning to end – at least, not to begin with. I read the Abstract if there is one, and then 'Conclusions', if there are any. Next I read the section called 'Discussion', next 'Results of research'; if no such sections are identified, I have to search a bit harder. I next scan the article for subheadings, then for graphs, tables and illustrations and their captions. The next procedure is a quick scan of the text itself for any keywords or phrases that match my profile (which does not need to be spelt out by me). Only when I have done all this – *if* I do it all, because I may decide to go no further at any stage – do I actually read any of the body of the text, and even then my reading may be confined to one or two sections or pages. Only a few articles are 'read' in any real sense; most are filtered out in the earlier stages, which is not to say that some information from them is not absorbed. It would be very interesting to know how others 'read' articles: self-observation over two or three weeks is needed to obtain an accurate picture of what people actually do, as opposed to what they think they do.

The whole process, at least up to the final reading, is a fast one. As a rough estimate, the abstract alone takes two or three minutes (it is short but usually dense); 'conclusions' between two and five minutes; 'results of research' between three and five minutes; scanning for subheadings etc. between one and five minutes (depending on the number and perceived interest of graphs, etc.); keyword scan five minutes – say ten to fifteen minutes on average up to this point (because few articles will permit all these procedures). The average time actually spent on an article is less, because although I read a few articles more or less in full I rarely go beyond the abstracts with many. The printed article is well adapted to speedy rejection – an inestimable virtue.

A well written article not only permits all or most of the procedures outlined above, it also reads as a narrative, with a beginning, a middle and an end. Its structure is a linear one, but because it is exposed on a few pages (typically ten or so), non-linear visual access is not only possible but easy, so that I can start with the end or the middle or do a rapid visual scan of the whole thing. The speed and ease of scanning for relevant items or passages should be stressed: if you consider how rapidly you can scan a newspaper, with its very large pages, you will appreciate the extent to which this kind of exposure aids access. If journal articles were printed on newspaper-size sheets, say one to a tabloid page, they could be scanned even faster.

The direct transfer to a screen of articles as they are produced at present creates no problems if the reader wants to read them through from beginning to end and if he has plenty of time to do so; his reading speed will be slower than in the case of the printed page, because the type is less legible and the screen holds only about half a page. If however he wants to follow my procedure, which I suggest is not untypical, he will find it very much slower and far less convenient, because it is less easy to override the linear structure on the screen. Means need therefore to be found of providing on the screen the facilities offered by the printed page. There are several ways in which this might be done.

1) Without altering the shape of the article, keyword scanning could be done automatically by a machine check of the text against a profile specified by the reader. This of course requires the reader to state his key terms, and works much better in subjects with a precise and standard terminology – science, technology and law – than in the humanities or, especially, the social sciences. It also runs the danger of missing relevant passages and/or of producing a lot of noise.
2) The order of sections in the article could be changed for display purposes so that the Abstract could be followed by the Conclusions, then the Discussion, then the Results of research. For this to be possible these sections need to be separable and identifiable.
3) A list of subheadings could be displayed immediately after the Abstract, and recalled if necessary after the Conclusions and Discussion. Sections could be called up as required. Obviously the article would have to be organised into clearly labelled sections.
4) Graphs, tables and illustrations could be displayed separately, preferably preceded by a list of such illustrative matter.

All of the above could be done without radical change in the form or structure of the article. What would be necessary would be careful labelling and coding of the text, which in turn would require a fairly tight or well organised structure, and a list of contents and illustrations. Such measures would go a fair way towards suiting my procedures. They would still be slower on the whole than use of the printed text, especially when it came to actually reading portions of the text (including the Conclusions etc.). Since all time would have to be paid for, the reader would feel himself under some pressure, which he could relieve by having specific tables or portions of text (for example) printed out for longer perusal.

The straight transfer of a document designed to be read in one medium to another medium is, however, unlikely to yield optimal results, and it should be asked whether the document can be specifically designed for on-line reading, taking full account of human needs and the limitations of the screen – 'limitations', because while an electronic system offers great advantages in fast retrieval of a particular article from a large corpus of documents, it seems to offer no advantages over the printed page once the article has been retrieved. In any redesign, it should be borne in mind that many articles will be printed out and read in that form, so that while the design should be specially suited to on-line reading it should also be suitable for 'normal' reading. More or less normal, that is, since ordinary computer printout bears a closer resemblance to a badly written scroll than to the familiar codex. There will undoubtedly be high quality printers which will produce something more acceptable, but these will be few and very expensive for some years to come and will not be directly available to most users.

The most appropriate design might be a pyramidal or tree structure (see Figure 1). At the top of the pyramid (or the

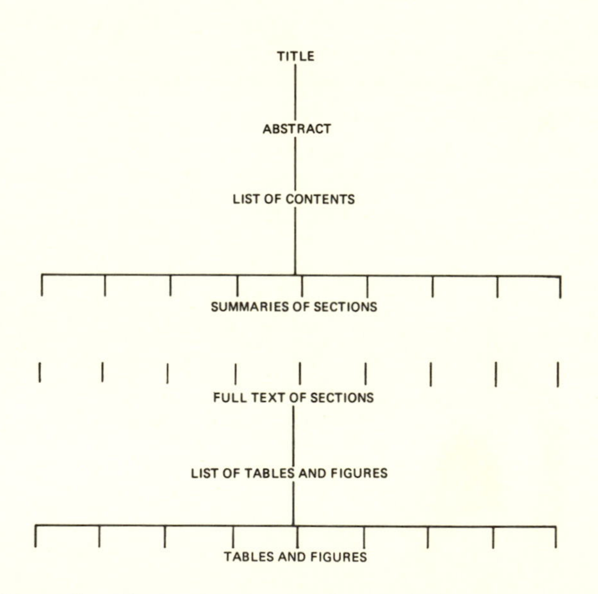

Figure 1. Hierarchy of exposure to machine-readable text

roots of the tree) would be the Title, possibly with keywords, as retrieved from the bibliographic data base. Next (the trunk?) would be the Abstract, the text of which would incidentally itself be searchable. This would be followed by the List of

Contents. Up to now the user has had no choice but to say 'stop' or 'carry on' or whatever the particular software tells him to do. Once he reaches the List of Contents he can select any of the items on it.

At this stage it would be most useful if he could call up a summary of any section, rather than the entire text of it. Even if a section is only two or three pages long, it is much easier to absorb its gist if its contents are reduced into a single paragraph. It is also easier to assess the relevance, and reject it if irrelevant: I have already suggested that assessment of relevance and significance (not the same) is something that could take excessive time with a poorly designed on-line system.

There would also be a List of Tables and Figures, from which individual tables or figures could be selected and called up.

There is nothing like practical experimentation for testing one's theories, so I set about redesigning one of my own papers along the lines suggested (one of my own because I would not have the temerity to redesign anyone else's). I chose a recent one that was relevant to the topics I have been considering, entitled 'Some questions concerning the unprinted word' (Article A). Immediately on looking at it again, I realised that it bore little resemblance to my archetypal scientific article (which in any case it does not pretend to be). It does not report research, it is not tightly structured, it is not divided into sections, it has no Abstract or Conclusions, and is not illustrated. Rather than turn to another article that conformed more closely to my model, I thought that to redesign it would be an excellent test of the concept, since if this could be effectively done there should be little difficulty with scientific articles.

The Title was not very specific and could well be missed by a retrieval system. I therefore provided a subtitle. Next I prepared an Abstract. I then had to divide the paper into sections that could be labelled. This proved fairly simple. I did not think it necessary to write Conclusions, since such a section would be little different from the Abstract. Since I had done no research, Discussion of results was not applicable. The final stage was to summarise each section. The sections were very unequal in length, ranging from a single short paragraph to two pages, but I nevertheless reduced them all. The results of this exercise are shown in Appendix

1, together with one page of the full paper for purposes of comparison.

This exercise made several things clear to me. The first is that if an article is not fairly well structured the task of redesigning would be very difficult. In this case the redesign took me no more than half to three-quarters of an hour. Conversely, if one has to design a paper in the first place for on-line reading it *has* to be well structured – a salutary discipline is imposed on the author. Secondly, once I had summarised each section I would probably find it quite hard to reconstitute the full text, as the Summary says most of what I had to say. The argument is omitted, and some of the Summary needs elucidation, but most of the article is there. When I used to do some abstracting many years ago I found that in the case of some articles the process was rather like boiling icecream or letting the air out of a balloon: what the author had to say was exceedingly little. Here again the design of papers for on-line viewing would expose the paucity of content of some articles, and for very few would recourse to full text be required. (This has incidentally been found in experiments with synopsis journals.)

The article I used for experiment was a very short one, with little 'solid' content. I therefore had another attempt with a different article of mine (Article B). This consists of twenty-one pages, has six tables and twelve figures, and is much closer to the 'typical' scientific paper in form and nature, although it lacks Conclusions. In this case only the Title, Abstract, List of Contents and List of Tables and Figures are shown, in Appendix 2. This article was easier to redesign; this confirms the (fairly obvious) suspicion that 'scientific' articles are more amenable to on-line viewing than articles presenting an argument, which are more common in the humanities and social sciences. Nevertheless, there is no reason why the latter should not also be, or be made, amenable to on-line reading.

It might be desirable to divide files of articles structured in the way suggested into two levels. The first level would consist of all but the full text, and should be adequate to satisfy at least 90 per cent of all needs. The second level would consist of the full text alone; since there would be relatively little call on this, it could be kept in a lower use store, possibly on microfiche coded for linkage with the higher level computer file, though videodisc would probably be more satisfactory. Note that the bibliographic references and abstracts would not be separated from the rest of the data in the upper level: in effect, the present files of references and abstracts would be supplemented by, or rather integrated with, additional data.

Articles structured as above could of course be printed out, and there could be a variety of printed versions for distribution: titles of articles, titles plus abstracts, titles plus lists of contents, titles plus summaries, and so on, all with full text available on demand. These need not be in standard packages intended for wide distribution like our present journals, though whether packages tailor-made for small bodies of users were economic would depend on how much the users would be prepared to pay, and the relative costs and benefits of the printed form and on-line access. It would be worthwhile to experiment with different forms of printed version to test acceptability and the market. For present purposes, the main points are that the on-line and printed forms could exist alongside one another, each having its own strengths, and that the printed version could benefit rather than suffer from being a sort of by-product of a tightly structured and summarised article intended primarily for on-line viewing.

There are however dangers in the tight structuring of articles. Writers might gradually drift into producing papers that were all structure and no style, and lose the facility of continuous writing. If this sounds unlikely, anyone who writes will be aware how long and possibily laborious a process it was to learn to write coherently, continuously, readably and with some sense of style. It is painfully clear that many writers, particularly of scientific papers, never achieve mastery of the art. These abilities are not innate, and in the absence of any incentive to develop them they may never be developed. It is worth also recalling how radio and television have changed over the last two or three decades; very rarely now do we see or hear the same face or voice for more than a minute, in case we become bored. It is assumed that we are all in need of constant variety and stimulus; seeming straightforward themes are fragmented, and narratives are presented as a series of small events in a different order from that in which they occurred.

Our span of attention must have been much reduced, but at the same time our ability to reconstruct from fragments has been increased. Future documentation may consist of masses of information which can be put together in various ways by the user, but are not packaged by the producer – 'producer' rather than 'author', because in this Lego-info world the makers of information bricks may no longer merit the latter title. Creativity will be as much the work of the consumer as of the producer. This might even be a good thing. Every benefit has its snag, and every snag its benefit.

To summarise. For on-line reading, articles need to be structured right and in a different way from printed articles in order to offer similar facilities of scanning and filtering. This can readily be done with the majority of papers, especially those reporting research findings. The discipline this imposes is good for the writer as well as the reader. From the machine version a variety of printed products could be produced, individually or for distribution. In such ways electronic technology would improve and extend access instead of restricting and reducing it. There could be important consequential effects of such changes on writing and reading.

References

1. This paper is a slightly revised version of one given to the Aslib Informatics Group at Oxford on 24 September 1981 and published in the proceedings of the conference (*Informatics 6: The Design of Information Systems for Human Beings*, edited by Kevin P. Jones and Heather Taylor, London, Aslib, 1981), from which it is reprinted with permission.
2. Line, Maurice B., 'On the design of information systems for human beings', *Aslib Proceedings*, 22 (7), July 1970, 320–35.
3. Lancaster, F. Wilfred, *Towards Paperless Information Systems*, New York, Academic Press, 1978.
4. Millson, R., 'The synopsis journal: its prospects for the scholarly author, publisher, and user', *Journal of Research Communication Studies*, 1 (4), October, 1979, 315–16.
5. Senders, John W., 'I have seen the future and it doesn't work: the electronic journal experiment', in Ethel G. Langlois (ed.), *Scholarly Publishing in an Era of Change*, Washington DC, Society for Scholarly Publishing, 1981, 8–9.

Appendix 1: Sample redesign of Article A:

Maurice B. Line, 'Some questions concerning the unprinted word', in Philip Hills (ed.) *The Future of the Printed Word*, London, Frances Pinter, 1980, p. 27–35.

TITLE

Some questions concerning the unprinted word.

Subtitle:

The scope and limitations of electronic transmission of documentary information.

ABSTRACT:

The radical changes offered by technological advances in information storage and access are not necessarily desirable for all forms of material. Electronic media have various economic, political, social and individual drawbacks, notably in their dependence on machines, which could hamper and restrict use. Technology should be used to extend rather than restrict the range of media available, and an effort should be made to match media to matter. The future system of storage and access should combine the unprinted and the printed word.

LIST OF CONTENTS

Features of a future electronic system
Economic factors as applied to different kinds of document
Technical factors
Social and political factors
The individual user
Factors determining suitability of different media
Existing printed material
A mixed future system?

SUMMARIES OF SECTIONS:

Features of a future electronic system
Preparation, editing and use will all be on-line. Microcomputers and microlibraries will supersede conventional libraries

and aid worldwide transmission. All forms of publication can be adapted to electronic modes.

Economic factors as applied to different kinds of document
Conventional publication is not necessarily uneconomic for most other forms of publication. Even for research papers the economics are not certain and alternative modes may be better.

Technical factors
Breakdowns and delays could hinder access.

Social and political factors
The machines needed for electronic access are not available to most of the world. Central control can be exercised over electronic media.

The individual user
Problems include the need to use machinery for electronic access, an unpleasant visual medium, and limited screen size. However, electronic media can be used to produce printed matter.

Factors determining suitability of different media
Relevant factors include the *nature* of material, its *audience*, and the *use* made of it. Categorisation by these factors can help decide what media are suited to what matter.

Existing printed material
There is a vast quantity of existing material that need not and should not be converted to electronic form except for conservation.

A mixed future system?
Printed forms should exist alongside electronic media, which will constitute the permanent form and from which they can be produced as necessary either in large numbers or on demand.

28 *Maurice B. Line*

As well as personal microcomputers and 'microlibraries', information will be accessible through television sets, which are especially well suited to pictorial matter and entertainment material such as plays and operas, as well as to the sort of information currently available through directories, encyclopedias and dictionaries. The coffee-table book, the reference work and the research paper will alike yield to the electronic revolution.

The effect on work will be great. Travel will become almost unnecessary. Conferences will take place on line; much office and administrative work will be done at home. Most of the manual work will of course be automated, so that management will become system management rather than man management.

All these things will be possible, but are they inevitable or desirable? Few would argue that none of them is inevitable or desirable. Economics will undoubtedly force some changes, and factors such as traffic congestion may well accelerate others such as the remote electronic office. More specifically, publication in conventional form is a grossly inappropriate, uneconomic and inefficient way of storing and making available research papers on topics of interest to perhaps ten or twenty people – and a considerable proportion of papers come within this category. The present procedures of searching bibliographic databases for relevant references and then having to request large numbers of papers on interlibrary loan, half of which prove on examination to be of no use for the purpose in question, suit no one.

At the other extreme there is no inevitability about popular novels, biographies or travel books. They continue to flourish, and to find a ready market even in difficult economic conditions. If they cease to be viable in conventional form, it is a safe bet that they will not be viable in electronic form. Nor is there any inevitability about textbooks, or even about many scholarly monographs, most of which presumably make a profit at present – this must be the assumption when new titles continue to appear at an undiminished rate. It is true that some publishers have been saying 'We have just managed to keep going up to now, but we are on the verge of collapse', but equally true that they have been saying this for some years. It is also true that some *have* collapsed, but publishing has always had a high failure

A page from 'Some Questions Concerning the Unprinted Word'

Appendix 2: Sample redesign of Article B:

Maurice B. Line, 'The structure of social science literature as shown by a large-scale citation analysis', *Social Science Information Studies*, 1 (2), 1981, 67–87.

TITLE:

The structure of social science literature as shown by a large-scale citation analysis.

ABSTRACT:

To collect information relevant to the improvement of secondary services, an analysis of 59,000 citations in the social sciences was conducted – 11,000 taken from 300 monographs, and 48,000 from 140 serials, including 47 highly cited titles and 47 taken at random. The analyses, the main results of which are summarised, included concentration and scatter, rank lists of cited serials, date distributions (crudely corrected for literature growth), subject relationships, country and language links, and forms of material cited. Large differences were apparent between analyses of references taken from serials and those from monographs, and smaller differences between analyses of references from highly cited and randomly chosen serials. This suggests that analyses based only on references from highly cited serials give a very incomplete picture. There were also great differences between subjects; most of these differences have implications for the planning of secondary services. The results imply that secondary services in the social sciences are deficient in the range of forms of material they cover, in the subject spread of material of possible relevance, and probably also of foreign language material and material published in many other countries.

LIST OF CONTENTS:

Introduction
Sources of references/citations
Concentration and scatter
Rank lists

Date distributions
Subject relationships
Country and language analyses
Forms of material cited
Implications for the collection of references for analysis
General comment
References (15)

TABLES:

1. Percentages of serials rarely cited, by subject
2. Serials in rank order of citations received from economics serial sources
3. Serials in rank order of citations received from psychology serial sources
4. Monograph authors (or editors) cited by serials 25 or more times
5. Annual serial citation decay factors, by subject
6. Annual citation decay factors, by form.

FIGURES:

1. Distribution of citations among serials compared with distribution of articles among serials
2. Percentages of serials in different subjects accounting for 50, 75 and 90 per cent of references made
3. Distribution of citations from serials among monograph titles and authors
4. Dates of references by serials and monographs to all subjects and forms of material
5. Proportion of references in different subjects, made by serials and monographs, pre-1964 and 1964–71
6. Proportion of references by serials and monographs to different forms of material, pre-1964 and 1964–71
7. Links between selected subjects as shown by serial references and citations
8. Subject 'inbreeding' as indicated by serial references made and citations received
9. References by serials and monographs beyond the social sciences
10. Links between areas of the world as shown by serial references and citations

11. Links between selected languages as shown by serial references and citations
12. References by serials and monographs to different forms of material.

4. INFORMATION TRANSFER IN EUROPE

JOHN MICHEL GIBB
Commission of the European Communities,
Luxembourg

The official title of this chapter is 'information transfer in Europe'. It is good to see the word 'Europe' in this title, but the presence of the word 'information' immediately raises problems for reasons which should be discussed before going any further.

What is information?

'Information' is a word derived from the verb 'to inform', and yet it seems to have developed a meaning, or rather a range of meanings, which have rendered it untrue to its origin.

The verb 'to inform' implies that a message of some sort has been issued, received and understood, i.e. that true communication has taken place. It has to do with an essential biological process (which is not the exclusive privilege of man) which can be split up into five basic steps:

1) an individual (or a set of individuals) holds knowledge of some sort;
2) he tries to express this knowledge in the form of a message made up of symbols;
3) he emits the message;
4) one or several recipients receive the message; and,
5) if all goes well, understand it.

The word 'information' has lost, or never had this wide connotation. It seems to refer essentially to the result of the first of the above steps, i.e. knowledge of some sort. Very often, it also covers the result of the second step, that is to say the rendering of the knowledge in the form of a written or spoken message.

This usage of the word makes it tempting to ignore vital phases of the process, such as making sure that the message is framed in a language which makes it accessible to those who can benefit from it, that it is actually emitted in such a way that the latter physically receive it and, with any luck, understand it. If the word were as rich in its connotation as the verb from which it is derived, the expression 'information transfer' would be pleonastic.

Another feature of the current usage of the word 'information' is that it is rather imprecise. It can refer to factual data, e.g. statistics, to bibliographical data, to the characteristics of certain devices or processes, to results of experiments, to more or less speculative considerations derived from these results, etc. This is not an undesirable feature, as long as it is recognised by those who use the word 'information'. (The situation in English is not as difficult as it is in French, where the word is used in the plural, 'informations', to refer sometimes to mere 'bits', handled by a computer.)

If one accepts this extremely wide connotation, it is not surprising that some should claim that about half of the economic activity of the more developed nations has to do with the origination, processing and dissemination of information.

The educational sector alone is an important component, which represents a larger part of the budgets of the developed countries. There is indeed no denying that education is a form of information transfer, in the widest sense, but it so happens that when talking of information transfer, one assumes that it is something different; at most the educational system has, as it were, prepared the ground for it. Clearly an individual requires a minimum of intellectual agility and background knowledge before he can become a viable recipient of information. Education is to information what a properly ploughed and fertilised field is to the seed that is sown into it.

All that has been said up to now is terribly obvious, but it is strange how often it is necessary to reiterate the obvious to avoid confusion.

Towards better information transfer in the research community

It is probably within the research community that information transfer was first consciously felt to be something of importance, that deserved study and needed to be improved as much as possible. The words were taken to refer to the dissemination of the results of research.

Information is seen here both as the product of research activities and a resource necessary to their successful deployment, which implies that information must be made to flow well within the research community if research work is to generate new information efficiently.

This has led to a tradition of scholarly publishing based essentially on journals. After the second world war, research boomed and gave rise, not unexpectedly, to a proliferation of journals and a few other carriers of primary information, such as conference proceedings. It was in those years that the expression 'information explosion' was coined.

Computerised databases – a partial solution

Thanks to reference publications of different kinds, in particular abstracts journals, it still proved possible to find one's way through the mass of primary literature that was being generated. In the 1960s these information retrieval tools began to be supplemented by computerised bibliographic databases. Much time and money has been devoted to improving these databases in terms of their ability to provide, in answer to queries, comprehensive lists of references to relevant documents. The development of telecommunications networks like Euronet has moreover made on-line access to them much easier, even over great distances.

Yet bibliographic databases have not met with the resounding success that some of the pioneers had predicted. On the other hand, data*banks* seem to be doing rather well, that is to say systems which provide actual information, for instance physical or chemical data, statistics, etc., rather than just indications as to where information may be found.

The reason for this is not that there is anything fundamentally wrong about the design and performance of the average database. The reasons must lie elsewhere. One of them, frequently cited, is that, whereas it is relatively easy today, with the help of databases, to detect interesting documents, obtaining physical access to them, and thereby to the information they contain, is often an elaborate affair and sometimes practically impossible. This is very frustrating for users. However, improvements in the organisation of document delivery, coupled with the introduction of new technologies, should provide ways out of this difficulty.

Applying the new information technologies

Those who expect miracles from the new technologies are, however, likely to be disappointed. They will in any case have to be patient (Gibb, 1979). It is customary, when talking of 'new' technologies in this context, to refer to the rapid progress which has been made in the last decade in the fields of data-processing, storage and transmission. It is not totally unrealistic to envisage the creation of huge stores of information (representing several years' production of hundreds of journals, for instance), the organisation of almost instantaneous remote access to them via telecommunications networks, the use of appropriate software to select material required and the rapid delivery of this material, even in large quantities (e.g. using satellites). This is something which has been, and is being, studied carefully within the European Community, notably by the Commission under its plans of action for information and documentation. It will be recalled that one of the first moves was to create the Euronet data transmission and switching network, which developed into the Euronet/DIANE system, namely the facility which now provides direct access, through Euronet, to the various databases and databanks 'spun' by the host computers situated at the nodes of the network. More recently, the problem of large-scale electronic storage and delivery of documents was investigated under the 'Artemis' study (Norman and Little, 1981).

This is not the place to comment in detail on the technological alternatives revealed by this study and their relative advantages in terms of effectiveness and cost. The larger scientific and technical publishers in Europe are obviously very much alert to the possibilities which are being opened up, seeing that their journals are produced via computer-aided photocomposition and are thus already available in a suitable form. They are encouraged no doubt by the steady developments on the other side of the Atlantic. At the same time, as large investments are involved, caution is necessary. An important consideration in any decisions to be made is obviously an estimate of the likely market for document delivery services.

The thesis which will be put forward here is that this market will not materialise truly as long as it is only passive. Let us not forget that the new technologies open up the possibility of generating and processing information on a decentralised basis. To use plain language, in the context of scientific and technical communication this means that an author himself could generate his original 'manuscript' in electronic form, say on a word processor, and if proper telecommunications facilities are available, transmit it directly to the editor of a journal, for instance.

A remarkable experiment was conducted by Professor Senders (Senders, 1981), of the University of Toronto, at the request of the National Science Foundation of the United States, covering the production of a totally electronic journal, one where all the traditional operations like typing the initial manuscript, sending it to referees, communication with referees, editing and distribution of the finished product were carried out using electronic means. The experiment did not work, for reasons which Professor Senders has explained in detail. As he put it at a conference: 'The experiment was a success, it was only the subject that died . . . The main reason was that a system must do what people want it to do or it will be a failure. In fact, there were so many constraints, it was so difficult to run, the terminal was not working when you wanted it, and so on.' In spite of this he remains optimistic and thinks the electronic journal is inevitable, as long as it becomes something very easy to operate. It is inevitable because the costs of conventional publishing will gradually become unbearable.

The conclusion to be derived from this is that there is a clear need for convergence between office automation, on the one hand, and data-processing/storage/transmission on the other. Even if the fully electronic journal concept, with all its complications, is set aside for a moment, it can be seen that there is much to be gained from bringing about a situation which will allow word processors to 'communicate' not only with photocomposers, but also with computers in general. Word processors would thus become also terminals through which electronic journals could be consulted and we could thus look forward to an era of integrated scientific and technical communication where authors, editors and publishers would work within the same system. Let us hope manufacturers will be sufficiently imaginative and aware of their long-term interests to bring this about.

The fundamental problem

This will not, however, provide the solution to a much more fundamental problem, which is fairly well known and yet receives relatively little attention. It is simply that many documents reporting on the results of research – this is true of journal articles in particular – serve the purpose, essentially, of recording the addition the author (or authors) of the document has made to the sum of human knowledge and providing evidence to the effect that it is a true addition. In other words, informing readers is not their primary role and, in any case, even if they are successful in this role, they are tailored principally to meet the needs of fellow specialists. This is not a criticism. It is impossible to write an article in such a way that it can be understood and appreciated equally well by a wide range of readers with different backgrounds.

There is no obvious way in which advanced technologies could be brought in to help solve this problem, which has to do with the intellectual process of making up the message that is to convey information.

Reviews

It is often claimed that articles of the review type can contribute to solving it, for instance critical reviews and state-of-the-art reviews, and that not enough of these are produced and published. The Commission ordered a study on this subject, which involved interviewing scientists and engineers. The study, which was completed toward the end of 1981, showed that reviews are indeed appreciated as means of seeing the wood, and not just the trees, both in one's own field and in neighbouring fields, but the results of the study indicate that scientists do not in fact consider the supply of review articles to be utterly unsatisfactory, though they would welcome an increase. In engineering and applied science in general, the situation is much less clear. Many of the engineers interviewed did not really seem to have an accurate perception of what review articles are, but when provided with an explanation came out firmly in their favour.

Unfortunately, whereas primary articles are plentiful, in view of the fact that they are one of the natural products of research, the same cannot be said of reviews, which are seldom produced on an author's own initiative. They are usually written at the request of a publisher or editor. It had been assumed, at the beginning of the study, that one of the factors likely to inhibit an adequate supply of reviews was reluctance on the part of authors – those best suited to write them being too busy managing research teams to find time to tackle such work. The picture that emerges from the data collected during the actual study is slightly different. It is generally regarded as an honour to be asked to write a review article, with the result that such requests tend to be accepted. The limiting factor would thus appear to be the number of requests issued by publishers or editors.

Synopses

Another device which can contribute to improving scientific and technical communication is the 'synoptic article' or 'synopsis'. A synopsis lies somewhere between an ordinary article and an abstract. Unlike an abstract, the essential role of which is namely to enable the reader to decide whether he is likely to find information of relevance to his interests in the article itself, and therefore whether it is worthwhile reading it, a synopsis is actually supposed to deliver information. It may contain illustrations and references as well as text. It is thus an article in miniature, which presents the essential results of a study or a piece of research, without enlarging, as in an ordinary article, on its links with previous work, on the equipment or methodology used, etc.

A number of synopsis journals have been created, offering their subscribers not only regular issues containing the actual synopses, but a back-up service providing copies, on request, of the full-scale articles or reports on which they are based. The

Journal of Chemical Research is probably the best known among those in Europe. It happens to be a European project in the widest sense, involving cooperation between learned societies established in different countries.

In spite of the attractiveness of the basic concept from the point of view of readers, these journals have encountered difficulties in their development, the greatest of which is 'author resistance'. It would appear that synopses, in spite of the fact that they are subjected to the same refereeing process as traditional articles and thus deserve an equivalent respectability, still tend to be regarded by authors as second-rate reflections of their achievements.

Perhaps the synopsis concept should be applied in a different way. This is what the Commission is trying to do at the moment in connection with the research programmes conducted under the ECSC (European Coal and Steel Community) treaty. The programmes cover a number of research contracts financed jointly by the private or public national bodies conducting the research and the Community. Contractual arrangements commit the Community's research partners into submitting final research reports, which are scrutinised by committees made up of experts from the member states before being released for publication. Inevitably, the reports are rather lengthy. Moreover, as most of them are highly specialised it is most difficult, not to say impossible, to direct copies of them precisely to the people who would derive benefit from them. Hence the decision to ask the research partners to produce, in addition, synopses of their reports. The collaboration had been secured beforehand of the editors of various European journals, specialised in coal or steel technology, who declared themselves prepared to publish these synopses (the full-length reports being of course available on request from the Commission). Thus the benefit of the scheme should be two-fold. It should, on the one hand, thanks to the role played by professional journals, bring these results to the attention of a wider public and, on the other, enable the reader to take in quickly the gist of the results of a research project.

At this stage, this scheme can only be considered as experimental. Time will tell whether the effort of producing and publishing synopses will have been worthwhile.

Overcoming the language barrier

The scheme has a special dimension which ought to be mentioned, namely multilingualism: the synopses will be published in the respective working languages of the collaborating journals. As for the full report, it is only available in the language of its author. Let us not forget that we are not dealing here with research in pure science, for which the natural common language for communication between European scientists has become English. On the contrary, we have to do with applied research in traditional industrial sectors which have deep cultural roots in the various countries in which they have developed and where research workers tend to report their results in their mother tongue.

If one excepts the relatively few people who communicate with complete ease in what has become the 'lingua franca' of science, all other Europeans are hampered in their efforts to communicate with each other by the language barrier. In the age of the 'information society', this presents a grouping of nations like the European Community with a tremendous challenge. It is not enough to ensure that more and more people have a better knowledge, at least a passive knowledge, of other languages, although that is by far the best solution to the problem. Aids to translation have to be devised. This explains why a relatively ambitious Community programme was launched in 1976. It covers a spectrum of sub-programmes ranging from the development of glossaries to full automatic translation.

Towards better information transfer in industry

In June 1981 the Commission of the European Communities held a symposium on 'The transfer and exploitation of scientific and technical information' (Gibb, 1981).

It was motivated by the realisation that the Community's future in a competitive world is conditioned by its capacity to innovate, and that innovation is impossible without satisfactory information transfer on as broad a scale as possible (and not only within the world of research). The Commission's aim was purely exploratory. Here are some of the terms of the call for papers it issued:

> How is the transfer and exploitation of information being effected today? More particularly, to what extent are the results of publicly funded research being brought to the attention of those in industry who are in the best position to exploit them?
>
> It is legitimate to suspect that many results remain unexploited, even if published. For example, the documents which present them are usually very scattered and written in such a way that they are not readily understood by those who are outside the research community. Improvements in the efficiency of transfer seem possible by developing the role of intermediaries.
>
> Who can contribute to improving the situation? The research workers themselves, journal editors, chambers of commerce, trade associations, documentation centres, technology transfer organisations, etc?

It will be seen that the Commission was less concerned here about information transfer between research workers, for which there exist ample traditions and media, than for effective transfer between research and industry. The response received was most gratifying. It proved impossible to fit in all the excellent papers that were submitted. The following points emerged very clearly from the presentations and discussions.

Information transfer, a problem of communication between people

First, information transfer is best achieved through personal contacts. Relatively few people in industry are 'information conscious', particularly in small and medium-sized enterprises. The representatives of advisory services, whether public or private, that approach such enterprises must do this with tact and understanding and ensure that whatever information is in due course provided is tailored to their customers' requirements.

In this sort of context two points can be made about computerised bibliographic databases. On the one hand, they can be seen as aids to ensuring, in a sense, the setting up of personal contacts. What else indeed are they than a means of leading up to communication between the author of a paper and his readers? On the other hand, if one bears in mind that it requires considerable skill, as things stand today, to extract satisfactory lists of references from these databases in terms both of recall and relevance and as, moreover, there then remains the task of consulting the documents that have been thus revealed and distilling out of them directly usable information, it becomes apparent that the average potential industrial beneficiary is not likely to take the trouble to consult databases directly. He may indeed prefer to use the services of an intermediary (information officer, information broker, etc).

Europeans are not communicating with each other

Second, it emerged that information transfer at a European level is practically non-existent, notably as far as the exploitation of patents and know-how is concerned. The institutions which cater for this kind of transfer are national and conform to strictly national requirements. This is not to say that they consider exchanges with other countries as immoral but, somehow, they work in an environment which inhibits international contacts. Even the flow of 'ordinary' non-proprietory information is impeded, if only because of the language barrier.

Third, if one looks across our European Community of something like 270 million people, it can be surmised that some interesting and useful initiatives are being implemented here and there which contribute to better information transfer. Are Europeans prepared to look at what other Europeans are doing in corners of the continent which happen not to be their own and learn from it?

The information market

Fourth, is information something that should be distributed free or should it be bought and sold on the marketplace? It was generally agreed that information has value and that its price should therefore be determined by normal market forces.

In actual practice, the situation is not so simple, for at least two reasons. First, the average consumer of information is not used to paying the full price for something which seems to be so immaterial. Let us look at the whole field of education for instance, which, as we saw above, is a major component of the information industry. For most people, education is something that costs nothing or very little – although, if they stop to think, they will realise that they are paying for education through their taxes. Second, and this is the most important point, awareness of the value of information is not widespread. An espionage agency is very much aware of its value and is prepared to spend large sums of money (and send people to their deaths) to secure information. Large firms maintain whole departments which do nothing but acquire and sift information, file patents, and sell licences, etc. At the other end of the scale, the smaller entrepreneurs tend to lead, in terms of their information requirements, a hand to mouth existence, perhaps mainly because they do not fully realise that they have such requirements.

Steps are being taken in some countries to redress this situation. France is an example, with its network of ARISTs. ARIST means 'Agence régionale d'information scientifique et technique'. There is an ARIST for most of the twenty-two regions which compose the country. The staff of these agencies approach local firms and offer them not information as such, but conduct a conversation in such a way as to detect what their problems are. In most cases, the solution to these problems does, of course, involve supplying information of some sort, as well as advice in general.

The level of charges for the services provided by the ARISTs are at the moment well below their real cost, but it is apparently hoped that 'information consciousness' in the smaller enterprises will gradually increase, which will in due course enable charges to be raised to more realistic levels. What is in effect being sought is an increase in the demand for information services, which should then develop not as a result of 'push' (technology push, for instance) but of market 'pull'.

Similar trends can be detected in other countries. It is being recognised that spending large sums of public money on information systems and services which are made available at heavily subsidised rates is, in the long run, the wrong thing to do, for a number of reasons. Launched by bureaucratic decisions, as is often the case, they do not necessarily meet real requirements or only do so inadequately. Moreover, they are sometimes also run bureaucratically, which is hardly conducive to efficiency.

It is also interesting to note that subsidising information services can be contrary to the free competition rules laid down in the EEC treaty. Let us take the example of two organisations offering European-wide (or, for that matter, world-wide) information services which are very similar. If one organisation operates along purely commercial lines, whereas the other receives a state subsidy and undercuts the former's prices, we have a situation which, if brought officially to the attention of the Commission of the European Communities, could lead to legal action against the second organisation.

Put more generally, the point to be made here is that the investment of too much public money in information services discourages private initiative and is ultimately self-defeating.

Does this mean that public money should be completely kept out of the information sector? Surely not, but perhaps public authorities should concentrate on pump-priming activities.

The need for intermediaries

A fifth point which was confirmed during the symposium and which grew naturally out of some of the other conclusions, is that effective information transfer supposes the existence of intermediaries. What different kinds of jobs are involved? Are there adequate training facilities for those who would like to take them up?

A first workshop was held by the Commission in October 1981 on the problem of training. It will lead to further European meetings, devoted each to the training of a particular category of intermediary, which will serve a two-fold purpose: first, enable those responsible for training in different countries to exchange their experience and, second, open up the possibility of giving the training courses a European dimension, that is, of providing those who follow them with a thorough insight into the advantages offered by the Common Market.

More radical measures are probably required in this field, designed to develop 'information consciousness' in general. In the societies in which we live, and this emerges from what has been said above, there are two sides to 'information consciousness'.

It implies, first, that as many people as possible are aware of the value of information, which should lead to an extensive demand for information services, and, secondly, that as many people as possible have an inkling of the skills (e.g. editing ability, knowledge of patent procedures) required to satisfy this demand, and of the aids which have been made available by modern technology, notably in the field of data-processing and telecommunications, etc., so that they possess the background permitting them to take up without too much difficulty work on the supply side of the equation. If this view is correct, some adjustments are required in university syllabuses.

The adjustments would involve, on the one hand, introducing into first degree science and engineering syllabuses elementary courses on information science and, on the other, setting up more facilities for intensive post-graduate education on that subject. Some already exist, although the emphasis still tends to be put at the moment on librarianship, i.e. on training a particular kind of information intermediary. This is not surprising. Librarianship is an 'old' profession in which there are well-established career prospects. The same cannot be said of information science in general although one can surmise, if it is true that 'information is power', that those who, thanks to their special training, are better than most at securing access to information and interpreting it, may ultimately qualify for senior managerial posts within their organisations.

What to do next?

The European Commission put foward its ideas about innovation to the European Council, the meeting of Heads of State or Government of the ten member countries, at Luxembourg in December 1980. In response, the European Council requested 'the competent authorities of the Community to examine ways of eliminating the fragmentation of markets and improving incentives to innovation and the dissemination of knowledge'. The Commission accordingly presented a policy paper in November 1981, which will be followed in due course by a number of detailed proposals, some of which will have to do with the promotion of information transfer. It is too early to predict accurately what projects will be involved. However, it is likely they will include the following.

An example: Holding more conferences on a European scale

No one seriously disputes that conferences are useful for information transfer, involving as they do both formal and informal communication. This explains why so many are being held. On the other hand, it is doubtful whether they are all as efficient as they could be. Perhaps giving a European dimension to more of the technical conferences that are being organised, could contribute to this.

Two broad types of conference are worth considering. One is the straightforward meeting which allows those active in development work in a particular technical field to get together and exchange their experience. Many of these meetings are held at a purely national level or, even if they claim to be a 'European conference on . . . ', fail to attract speakers and participants who are representative of the expertise available throughout the continent, let alone adequate attendance from other parts of the world. It is of course very difficult to organise a true European conference, with balanced participation from all the countries that could usefully contribute. It is not enough to provide simultaneous interpretation at the conference itself. The main problems occur at a much earlier stage: establishing the necessary contacts to secure the participation of suitable keynote speakers, producing calls for papers, programmes etc., in several languages, locating the various national media through which to advertise the conference, and so on. The Commission therefore intends to launch a scheme whereby the expertise which it happens to possess in this field, together with some financial incentives, could be offered to those organisations which are prepared to face these difficulties, notably on the basis of trans-European collaboration.

Similar incentives would be provided for conferences of the 'current awareness' or 'state of the art' type, the objective of which is not to allow experts to exchange views, but rather to communicate to a wider audience the opportunities opened up by certain maturing technologies. The initiative of such conferences is often taken by publishers, who regard them as a means of putting together a useful book, which they proceed to offer to an even wider public. Initiatives of this sort are usually taken at a national level, most of the speakers/authors being from one country only, and lead up to a monolingual publication. It has occurred to the Commission that the publication could be of higher quality and have a greater impact if it were based on a conference organised at a European level and issued in several languages.

These are only two examples of the many measures that are being envisaged to improve information transfer in Europe. They would not make very large demands on the budget of the European Community, but this does not mean that their impact would necessarily be small. The states that have joined the European Community have done so basically because it makes good sense, in many sectors, to work together. If it is true that information transfer is one of these sectors, it should not be necessary to spend large quantities of money to enforce the collaboration that it requires, but should be enough simply to help common sense to bring it about.

Conclusion

It would be pretentious to try and be conclusive about the question of information transfer in Europe. The subject is too vast and needs much further exploration. The only conclusion that will be hazarded here is that the major problems involved are educational, organisational, psychological and social. Thus, although technology can contribute to solving them, the real solutions lie elsewhere and involve, as in many other areas of human activity, common sense and sound management.

References

Gibb, J. M. (ed.) (1979), 'The impact of new technologies on publishing', *Proceedings* of the symposium held by the Commission of the European Communities in Luxembourg, 6–7 November, Munich, K. G. Saur.

Gibb, J. M. (ed.) (1981), 'The transfer and exploitation of scientific and technical information', *Proceedings* of the symposium held by the Commission of the European Communities in Luxembourg, 10–12 June

(EUR 7716 EN), Luxembourg, Office for Official Publications of the European Communities.

Norman, A. and Little, Arthur D. (1981), 'Electronic document delivery. The ARTEMIS concept for document digitalisation and teletransmission', Oxford, Learned Information.

Senders, J. W. (1981), 'The electronic journal', *EURIM 4 Proceedings*, London, Aslib.

5. VIDEOTEX IN THE UK: PROBLEMS OF PUBLIC SERVICE VIEWDATA AND IMPLICATIONS FOR PUBLISHERS[1]

BARRY J. WITCHER
Videotex Marketing Applications,
Department of Marketing,
University of Strathclyde, Glasgow, UK

A videotex family of definitions and terms

Public service videotex began in the United Kingdom in the late 1970s. The services were Prestel, British Telecom's[2] public viewdata service, and Ceefax and Oracle, British television's teletext[3] services. It will help exposition if these, and some other terms are explained. They are outlined in Figure 1.

Videotex is the generic name for the group of electronic communication systems which make use of television screens to display computer-based information. Its main difference from ordinary computer systems using visual display units is in data presentation and retrieval. Information is displayed in page form, in single pictures of text and simple graphics.

Information is obtained simply by pressing the numbers of pages on a hand held keypad. No prior training is necessary. The videotex user merely needs to know the number of the page on which the required information is located. Numbers can be looked up in printed directories, or indexed on the videotex system itself. Alternatively, the user may browse through system pages, by finding a way through the series of guides, designed to route readers to the information required.

Editorial skills are required on the part of videotex information providers (IPs) to present, lay out and index the information service, much as might be done for more conventional publishing media. This is unlike computer information retrieval systems which call up data in units or bits, usually through command languages and trained intermediaries.

Videotex is designed for general readership. It is in this sense a publishing medium of communication rather than a computer one.

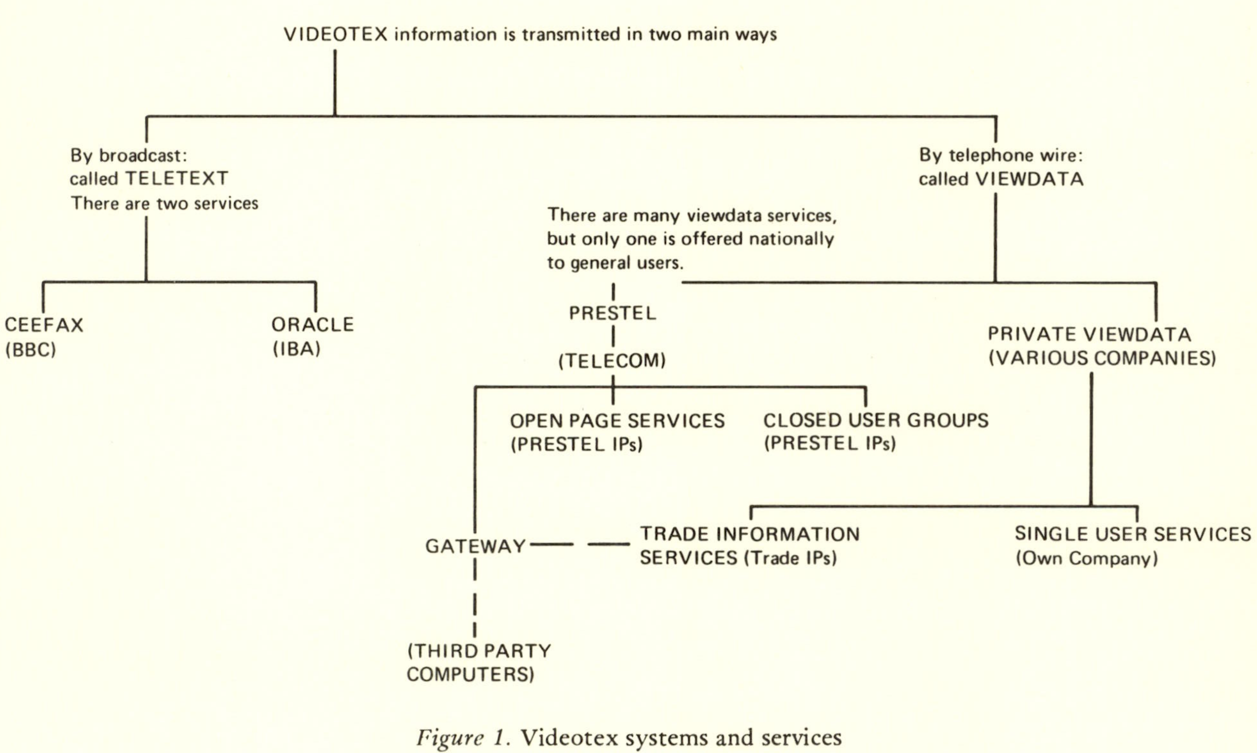

Figure 1. Videotex systems and services

Public service videotex or viewdata refers to a national availability for general public use. A subscriber need only have a modified television set or computer terminal to link into the service. Sometimes videotex is called a mass medium. But this is only true in terms of availability. There are important differences to other mass media.

These are user selectivity and the interactive nature of some forms of videotex. Information is called up by users when they require it, and usually in the most up-to-date form. This may be acted upon immediately through the medium, either to send a message to an IP, or request further information, reserve or buy something.

This cannot be done with other mass media, and is difficult even with a combination of, say, newspaper and telephone. Much of mass media information is general and too undirected to be of use for specific purposes, and follow-up action is usually required.

Not all videotex services are equally handy, however. Figure 1 shows videotex as divided into two sub-groups, depending on how information is relayed from its computer source. These are called teletext and viewdata. Teletext receives its information in a broadcast form over the air, in the same way that ordinary television programmes are sent. There are two services received in this way. Ceefax, which is supplied by the British Broadcasting Corporation, is one; the other is Oracle, the teletext service provided by the Independent Broadcasting Authority.

Teletext is a one-way communication system and is not interactive. Though teletext received by cable television can be made so. Some experiments have been carried out in the United States, where it is often termed interactive teletext.

Viewdata receives information over telephone wires, and is an interactive medium. For this reason it is sometimes known as interactive videotex. Two sorts of viewdata are usually distinguished in the literature. These are Prestel or public service viewdata, and private viewdata, which is organised and provided by companies other than Telecom (or in other countries, by companies without links to national postal, telegraph and telephone authorities).

Not all the information contained on Prestel pages is available for unrestricted public use. Those pages which are confidential to IPs and authorised users may be hidden or unlisted in directories, or held within closed user groups (CUGs). CUGs are usually organised by IPs for users willing to pay subscriptions, who in return are given codes and access to the information service. Otherwise Prestel pages are open to access by anyone. Payment is made on a pay-as-you-go basis.

There are two kinds of private viewdata service. One is similar to the idea of Prestel CUGs: a dominant IP or organiser provides a coordinated service for users in a particular trade or business. Users pay a subscription, and in return are provided with equipment, and access to the IP's storage computer. The information comes from outside user company organisations but is necessary to its activities.

The other kind of private viewdata is usually provided by an equipment company or software house for use within a single organisation. The information is generally generated internally, and an important feature of the system is that company personnel find it as easy to use for putting in information as for extracting it. In-house private viewdata is thus an extension for the layman of the company's computer, with electronic extras such as company magazines and messaging.

Prestel is a technically underdeveloped system when compared to most private viewdata systems. Numeric keypads make messaging difficult. Messages have in any case to be left with the central Telecom computers for IPs to collect. Thus, strictly speaking, Prestel is not interactive, nor real time. There can be a delay while the IP picks up user messages and replies. Putting information into the system is also time-consuming for the IP.

However, at the present time (April, 1982) a user to user message system is being tested on the network, and later 'Gateway' is to be introduced. This is a facility which will allow users of Prestel to link up with computers, and perhaps private viewdata, outside Telecom's storage computers. If this works perfectly, it should permit interactive use, and in real time, of third party databanks. It should also greatly increase the information available to the Prestel user, and allow the use of keywords and other searching techniques.

The Prestel concept

Prestel was seen by many in the early days as a kind of information warehouse. Quite unlike teletext or private viewdata Prestel was, and remains, a conglomerate information medium, where the service was not organised or controlled by a single editor, but was in effect, a collection of independent services. Telecom adopted a common carrier policy, where content would be left to the IPs, and computer space was rented out to interested parties on a first come, first served basis.

There are about 800 organisations which supply services on Prestel. Some 179 of these are main IPs who hold contracts directly with Telecom: this costs £5,000 and £5 a page, per year. Many of these IPs act as umbrella consultants, sub-letting pages to the other 655 IPs, and giving editorial assistance. This can cost a sub-IP about £50 a page, per year. A CUG can be formed for an extra £250 a year, over and above the £5,000.

IP pages currently amount to around 217,000. This is large by teletext standards where pages are limited to 500–700, and private viewdata, where pages are to be thought of in hundreds rather than thousands.

The idea of a warehouse for information lay in the old goal of a computer for everyman. Sam Fedida, the Telecom 'father of viewdata', and Rex Malik, have written that the original intention for Prestel had been to '. . . create an information retrieval system for mass market use, one which would get around the limitations of existing computer systems (complex, skill-demanding, expensive), which excluded them from the mass market . . . had to be aimed at millions of users, not the thousands which are the norm, and with which the computer industry and community are familiar and at ease' (1979, p. 6).

The consequences of achieving this would be to bring forward the dazzling prospects of electronic futures and socio-economic change. Viewdata in the third revolution (of electronic change), would do what the steam engine did in the second (industrial revolution), and turnips did for the first (agricultural revolution). And for a while, Prestel's champions thought they had done it: '[Prestel is] a, major new medium . . . comparable with print, radio, and television . . . [leading to] major changes in social habits and styles of life, and [will] have long-lasting as well as complex economic effects.' (op. cit., p. 1).

No wonder many of the first IPs to come forward for Prestel were from traditional media backgrounds, particularly publishing. It was clearly a new medium to be concerned with, and, perhaps feared.

The main possibilities for traditional publishing are outlined in Table 1, and are of three main types:

Table 1. Electronic consequences for traditional publishing media

DISPLACEMENT through:	transfer of advertising revenue; user pull replacing information push mechanisms; more effective media for fast and convenient information.
COMPLEMENTARY through:	management and marketing tools for publishing; updating medium for existing media; indepth, individual user follow-up service for existing media.
NEW OPPORTUNITIES through:	information tele-propinquity and possibilities for service/product evaluations, source and information broking, problem solving.

The displacement effect

The greatest danger lies in the possibility that some publications would be forced to cease altogether as competition for advertising revenue moves in favour of videotex media. This thesis is supported by advertising trends for media revenue throughout the 1970s.[4] Revenue in real value terms has been fairly constant. If this were to continue, and videotex were to become more commercially important, then advertising revenue would be taken from traditional media forms.

It may be argued that the swing away from printed media advertising to electronic forms has already been in progress for the past twenty years or more. Television and radio have taken a greater share of the constant media revenue cake.

In addition, videotex may do things much better than the traditional forms. Specifically, it will satisfy the need for quick, convenient, and up-to-date information like that provided by directory and yellow pages or by classified advertisements and mail order. Videotex is a near perfect medium for direct marketing and selling.

Directory, classified media, and direct marketing, are taking a growing share of advertising expenditure in traditional media.[5] They are likely to increase and transfer to electronic media if they prove more effective. Viewdata has the capacity to update selling and advertising information almost immediately changes become known. There need be no directory problems of datedness, and geographical incompleteness of coverage.

More generally the greater effectiveness of viewdata is likely to be felt in a pervasive way. A curse of our times is information overload, or the piling up of reports, magazines, papers and so on. This may be most true of technical and managerial fields, but it is also true to an extent, for consumers, particularly durables. Product and service advertising abound, and libraries, consumer advisory groups can only counter with more reports and leaflets. Only a small proportion of the paper received will actually contain wanted information, the rest being too general and irrelevant.

This is essentially a problem of information push. Viewdata offers the potential of the user pull. Where the user can individually select the material he needs from its primary source. In the long run, this is likely to have a major contracting effect on traditional publishing as push are replaced by pull mechanisms in information dissemination and extension.

Complementary applications

Viewdata may be useful to traditional publishing simply as a useful aid to promoting sales, improving links with company distributors and customers. Both can be kept informed about latest stock and delivery information, and reviews. The publisher could monitor advertising promotion effects, by observing and linking viewdata page use with information.

As part of multi-media campaigns, viewdata should be seen as an extension of old forms of publishing. To provide more, perhaps factual, information on products and services advertised in traditional media. All a conventional advertisement need do is include a viewdata page number at its foot to refer the interested party to further details.

Books and directories could be updated in a similar way. There is already one book in which readers are invited to view updates on Prestel pages. It is, of course, a book about Prestel (Nicholson and Consterdine, 1980).

Another kind of multi-application for old and new media might be used for news. Electronic newspaper readers could be provided with a facility enabling them to request follow-up analysis of interesting news stories, or a collection of previous stories concerning a news item or subject of interest. This may best be supplied as conventionally printed material, or even by radio/cable television.

New opportunities

There are also likely to be completely new roles for media and publishers. The most likely will emerge from opportunities in evaluative and consumer advisory services. These are likely to include new ways of marketing products and services, and more testing procedures. There are likely to be new roles for publishing groups as information system organisers and information brokers, as final consumers move closer to primary data sources, and computer system designers tailor functions to business needs.

Information telepropinquity or the closeness of information to transactional activity in the home is as yet little understood for its importance to retailing. Conventional wisdom has it that electronic shopping and marketing will transform convenience and routine shopping. However this is to ignore the importance of information to consumer search and conditions of imperfect competition for consumer durables and professional services. Viewdata is likely to have a greater impact in the specialist, non-routine shopping sector, than it is for convenience trade.[6] And with the changes in trading systems will come an enhanced role for IPs.

But these are events of the future. The problem is to get there. Visions and speculations rarely turn out in quite the form they are expected. This is particularly true of major new technologies. The road to commercial gain is littered with the disappointment of product champions.

Prestel's progress

Prestel came about because technological advances in micro-electronics and telecommunications enabled the vision of individuals, such as Fedida, to become commercial possibilities. Development was encouraged by parallel developments overseas and in television's teletext. But there was no obvious market demand for electronic home information services pulling view-data into existence; it was therefore, a technology push innovation, and a very radical one at that.

Prestel's progress has been slower than expected. At the start of the national extension of the public service in the autumn of 1979, around 1,700 Prestel registrations were recorded. Telecom predicted 'tens of thousands of customers . . .' by the end of 1980, which would be the basis for growth to millions in the 1980s (Telecom, 1980).

Early in 1981 the National Economic Development Office were forecasting 46,000 registrations at the year's end, and a million by 1985 (*Viewdata and TV User*, 1981). By March 1982 total registrations were 13,933. Monthly totals seemed to have settled at a growth rate of 400 to 500 a month.

The forecasts, many of them based on comparisons with the early market development phase for television, now seem silly. Existing habits and life styles are based on not having electronic information media. Television was a natural extension to radio. There is no obvious precursor to an electronic information facility in the home.

The future of mass market viewdata seems postponed for the moment. A historical outline is given in Table 2.

Events seem to have taken the Prestel concept through two, and perhaps now, into a third stage. From a residential, basically publishing medium, to a business oriented service, and with the advent of Gateway, perhaps to a computer network for third party databases. In the final form, the idea of central storage computers may come, gradually, to seem like an expensive white elephant to Telecom, a company whose traditional business has been of network manager rather than a keeper of information.

If Prestel were to become only a service network for third party databanks, then the difference between original Mark I

Table 2. Prestel's phases and changing concepts

Early 1970s:	TECHNOLOGY DEVELOPMENT STAGE
	Major influences were – viewphone, the layman's computer/using the TV screen/teletext display methods
	Prestel Concept Mark I: A cheap, pay as you go, easy to use information retrievable system for the mass residential market.
	A publishing medium and extension to the TV set. Conventional office and computer interests not given prominence.
Late 1970s:	MARKET DEVELOPMENT AND COMMERCIALISATION STAGE
October 1976	First IP, Consumers' Association.
June 1978	First user terminal in home.
October	Market trial, later called test service, began.
March 1979	Public service opened. 694 registrations.
September	Service extended outside London. 1400 registrations.
March 1980	National promotion of Prestel concept. 2800 registrations.
Early 1980s:	ADAPTATION STAGE
November	Changed marketing strategy favouring business markets, announced to IPs. 7000 registrations.
	Prestel Concept Mark II. A pay as you go service, but also with CUGs involving subscriptions, and general service more targeted to business markets.
	Still essentially a publishing medium but important role for terminals and business equipment suppliers.
September 1981	Network call capacity reduced and Prestel centralised. 11500 registrations.
October	Mailbox introduced.
1982	Gateway and picture Prestel.
	Prestel Concept Mark III? A high cost network, linking private viewdata, on-line computer services, for business and specialist subscribers.
	Business needs and keyword developments make service less of a publishing medium and more of a business computer system network. Major role for computer, including personal computer companies.

Prestel and the final Mark III (Table 2) is very great. It is the difference between a far-flung social network, with maximum access for all citizens, and one which offers only privileged access to private and specialised databases. The difference in terms of social service could be immense.

The need for social service viewdata is more obvious than its economic viability. This certainly applies to the early years when public subsidy is likely to be required.[7] Without subsidy, only the high socio-economic groups would benefit. Nor is a public service likely to be widespread without consumer education and the time this takes: people have to be led into a change of habits.

The problems associated with Prestel's market development are briefly set out in Table 3.

Table 3. Prestel's problems

Two main factors were inherent weaknesses to the Prestel programme:

(1) radical, technology push nature of the innovation:

(2) a strong faith in a mass market for electronic information.

This created three handicaps for later on:

(1) participants too dependent on the creation of a mass market, and not used to market development:

(2) neglect of concept testing, marketing procedures, and coordination of services and products:

(3) poor product and service design.

These things made it difficult to later adapt to mistakes and to changes in policy.

Conventional view: costs too high to establish a mass market, but market too small to bring down costs.

The often quoted chicken and egg problem of high costs/no market–no market/high costs, has been overstated. If equipment had cost much less it is likely that other costs would have proved a sufficient deterrent.[8] It is not so certain that these would have come down with larger markets. Particularly if consumers' perceptions of information differences between Prestel and teletext are not marked, teletext's virtually free use of information must appeal.

The chicken and egg problem is really a symptom of poor product design and marketing to begin with (see Table 3). The potential of Prestel as an important domestic facility is in its interactive qualities, and these have yet to be developed. Until they are, Prestel IPs are likely to go feeling the chilly winds of tight budgets.

Most of the present IPs have been with Prestel from the beginning when concessionary charges for IP contracts were available. Their reasons for involvement were various, but there seems to have been a common realisation that an electronic era of communications lay round the corner. Their objective was to establish a foothold from which to develop an expertise to capitalise upon market opportunities should Prestel be successful.

This has dangers for the development of the service where 'footholds' have not materialised into top management commitment and adequate financing. IP running costs are high and very few, if any, IPs are making profits from the provision of information alone. A large IP, with more than 1,000 pages, might find costs approaching £50 a page, depending on the size of overheads. To be running a profitable database from pages alone, a large number of accesses will be required. As a result there are only a handful of IPs employing large staffs; most are composed of a staff of only a few individuals who are on a shoestring budget.

The quality implications for the Prestel database are omissions, superficial and misleading information, some of it out of date. The result is that the service gives a ragged impression to the user, which in some instances is made worse by confused indexing.

IPs have been able to agree upon a voluntary code of conduct. This is published by the Association of Viewdata Information Providers and determines the ground rules for information and advertising design and how these relate to advertising and legal constraints.

Marketing strategy

Until late 1980 Prestel's marketing had centred on the promotion of Prestel as a concept, with a general, residential market. But at the end of November 1980 only 12 per cent of registered Prestel sets were in homes. And it is likely that some of these were for business purposes. The one significant group of users was the travel industry, mainly travel agents. This group accounted for 19 per cent of the overall total of registrations. The next largest groups were financial, and computing/electronics, both at 3 per cent.

Given the disappointing residential market penetration, marketing policy changed, to direct resources at the business community, and within that, a number of key segments. These were chosen as areas for which Prestel already provided a nucleus of information services, and which might be considered able to bear the high costs. The domestic market would not be abandoned, but left for a time while equipment prices come down. The target segments, upon which most of Telecom promotion would concentrate for 1981 and 1982 are given in Table 4.

The change of marketing policy was of concern to those interests most dependent upon the development of Prestel for a mass residential market. These included the manufacturers of television sets and general database IPs, particularly those that provide social information.

Buying decisions for Prestel in business are likely to be associated with information needs. This contrasts with decisions in the residential market, where attention is concentrated on the television set; this will usually be in a rebuy context, where Prestel is essentially an extra rather than a main reason for a purchase. The business market offers, therefore, more scope to manufacturers of terminals rather than receivers. Many of these will have their main interests in the office equipment and computer accessory markets, and should prove more flexible than the mass market manufacturers to particular business needs.

The importance of information quality to the terminal purchase decision has persuaded Telecom to take a greater interventionist stance on information provision than hitherto.

Table 4. Target market segments – progress to March, 1982.

Market	Size	1982 Target (per cent)	August 1981 (per cent)	Prestel registration share (per cent)
Travel	5,000	100	49	18
Financial*	30,000	27	2**	5**
Commercial property	2,000	50	9	
Agriculture	10,000	15	1	1
Construction	15,000	13	1	1
Hotels	34,000	2	0.2	0.5
Lawyers	7,500	7	0.5	0.3

Notes: *Financial includes commodities, banking, insurance users, etc.

Note of explanation

'1982 Target' shows the degree of market penetration aimed for by end-1982.

'March 1982' refers to the degree of penetration achieved by March 1982.

'Prestel Registration Share' is the share of total registrations taken by the market sectors, end-March 1982.

Source: Estimated from Prestel statistics.

Late in 1980 Telecom presented its marketing strategy to Prestel IPs at a conference. Here it declared its view that information sells sets, and thus IPs should review their databases to ensure they have a product which can be marketed to, and thus wanted by, the business segments Telecom had specified with potential for market growth.

Telecom would in future encourage those IPs with good quality and suitable services. Those with a poor record of incomplete and outdated pages might not have their contracts renewed in the future. Pages given to Prestel for indexing might be ignored if they seem unsuitable or likely to bring the service into disrepute. It would also become more difficult to become a new IP on Prestel.

It is questionable how IPs have been able to tailor their databases to specific sector needs. There are few instances of pages which actually provide information useful to the day-to-day management of industry. Ideally, IPs need to buy marketing and selling expertise in those areas in which they hope to provide information, pilot services and place sets in user environments. This should be done with the co-operation of equipment suppliers. The rub occurs when it is realised that subsidy and top management commitment is usually required.

It can be done, as the IP Sealink has proved. This is a ferry company which has worked closely with travel agent groups in the general viewdata area. Since October 1979 it has taken the lead and operated a partnership with Granada TV Rentals Limited to carry out a scheme to subsidise agents' Prestel costs. Close to 90,000 accesses a month are made to the Sealink database – which is one of the most complete on Prestel, covering nearly all the questions that can be asked about Sealink. The present financial commitment is £500,000 over two years, but if one in a thousand accesses results in an extra booking, Sealink estimates that the annual income, at the present number of Prestel users, will be about £100,000.[9]

The success of Prestel in the travel industry owes much to the Sealink initiative. This is reflected in lists of the more popular Prestel IPs. These are shown in Table 5. Five of the IPs listed in Table 5a) and seven in 5b) are travel or holiday organisations, and some of the others have travel information as a prominent part of their databases.

Business IP services are important in 5b). The Intercom Videotex and Stock Exchange databases consist of relatively few pages but these are updated regularly: thus users may use a given page several times over a short period. Most of the information is about commodity and stock prices. While this is specialised information it seems to be accessed by a general business audience. This is the kind of use for which Prestel is probably most suitable, in other words, for the general business user, whose use of specialised information is too occasional to justify expensive subscriptions to more sophisticated services, but regular enough to require the latest information.

The problem for Prestel is how to create this kind of specialised service, when more profitable markets for the specialised IP are probably in more sophisticated markets, where subscriptions and higher usage rates give greater guarantee of return. Of course, if the marketing strategy leads to the creation of specialist

Table 5. Leading Information Providers during August, 1981

a) By number of times a particular service is used:

IP's name	Nature of service content
*1 Viewtel 202 (*Birmingham Post*)	Self-styled 'world's first electronic newspaper'. News items and advertising.
2. ABTA	Travel information guide.
*3. Mills and Allen	Magazine subjects, including games.
*4. Baric	Magazine subjects, including travel.
5. Thomson Holidays	Range of holidays.
*6. Fintel (*Financial Times*)	Business information.
7. Sealink	Ferry company.
*8. AVS Intext	Business, jobs, etc.
9. Thomas Cook	Travel agency.
10. British Airways	Travel.

b) By number of times pages are used:

IP's name	Nature of service content
1. Intercom Videotex	Commodity prices.
2. Stock Exchange	Financial.
3. ABTA	
4. Pan American	Travel.
5. Thomson Holidays	
6. Quantas	Travel.
7. Cosmos	Travel, holidays.
8. Horizon	Holidays.
9. British Caledonian	Travel.
10. Viewtel 202	

Note: *denotes umbrella IPs

Source: Prestel

services and users on Prestel, there may then be a large enough market overall to warrant specialised services for more occasional users. This may take a long time. The present penetration rates for target segments seem low if 1982 targets are going to be achieved (see Table 4).

Electronic newspaper

The most popular database on Prestel in terms of number of accesses is Viewtel 202 of the *Birmingham Post*. Viewtel styles itself the 'world's first electronic newspaper'.[10] It carries articles and magazine-type items, much as a newspaper would do, though perhaps the space is less on the topics covered – a Prestel page carries about two paragraphs of an ordinary newspaper. News is updated on a regular basis. In May 1981 Viewtel accounted for 11 per cent of Prestel accesses, while having only 2 per cent of pages.

Pages are free to view since most of the costs of the service are met from advertising. This includes directory advertising, including events and yellow page-type material; display, using videotex 'adflashing' to catch the eye and refer users to other pages, and classified advertising.

Viewtel's success and experience has encouraged it to consider a private viewdata service for the West Midlands, an area of two million households and businesses. Suspecting that viewdata will grow largely outside Prestel, they have thought of using a subscription service based on Telecom's public packet switched service, a new system of transmitting computer information in packets of bits over the telephone network. Given a reasonable market, the annual cost to a subscriber would be less than for a quality newspaper.

The future

The possibility that Viewtel might expand outside Prestel is perhaps a warning for the future of public service viewdata. It may indicate that IPs will increasingly place their efforts elsewhere, while only marking time on Prestel. The possibility of Prestel III as a network system service has already been indicated. It is particularly likely to materialise if the open page information on Prestel's central computers becomes too costly to maintain depth and quality, while on the other hand, CUGs and Gateway become successful.

Everything about videotex is still in a state of flux. One of the most important questions is how expensive will Gateway be, and how will it compare to the costs of packet switched viewdata. If Gateway costs are low, and cost comparisons are favourable to Prestel, then it is possible that new uses for public service viewdata will come about.

As originally envisaged, as a medium of communication for residential markets, Prestel would have had its impact on publications of perishable and fast information, where compact presentation was involved. Thus newsletters, guides, advisory material, directories and timetables, and classified advertisements, were the most likely type of publishing to be affected. It seemed likely that newspapers might have most to lose in the long run.

It was thought that other published material – any that is descriptive or lengthy, for which demand is likely to be occasional and specific, and which the user goes straight to, rather than scans for items of interest – would not be in competition. This is no longer so. The Gateway facility will permit the Prestel user to access specialised third party databases and existing on-line services. Archival and other bulk information may become available.

The change in demand for published material is uncertain. What may happen is that whereas most information is pushed at potential users, videotex and other developments will create pull mechanisms, where the user requests specifically what he wants from the primary source. It could mean a major streamlining of abstracting and indexing services, as the flood of primary literature, particularly in selling and technical advice, declines along with the need for attracting users and providing hard print reference material.

But before this happens, there is likely to be a transition, during which videotex networks remain small-scale and access to databases is limited. In this period networks' main use may be as sourcing media, where IPs act as electronic clearing houses which direct users to products, services and literature for further reference and more detailed inspection. The IP is in effect acting as a middleman to help the user find from traditional sources the service or information he requires.

The demand, however, seems likely to come from business, and will depend for some time ahead on the level of subscription charges laid down by the IPs. Ready access to third party databases will not be widely available to Prestel users without prior arrangements with the IPs or CUGs.

Towards the end of the transition period, perhaps effectively ending it during the early 1990s, the development of transactional videotex may see Prestel IV, when teleshopping and associated tele-information services will begin to exploit residential markets, and result in a broadening of use for the Prestel network.

It is certain that the active participation of publishers in ventures such as Viewtel will stand them in good stead and be of great value in the times of change, which still wait around the corner. This is so, even if experience with public service viewdata indicates that those changes are coming later than was once suggested.

Notes

1. This article is an outcome of the Prestel Innovation and Marketing Strategy Research Project which is funded by the Science/Social Science Research Councils' Joint Committee.
2. British Telecom is the new name for Post Office Telecommunications. The company no longer has any connection with the British Post Office. However, it remains the UK's central authority for the supply and organisation of the public telephone network. Throughout the text the company will be referred to as Telecom.
3. Teletext is not to be confused with Teletex, which refers to the system of transmitting text and documents between terminals, and has no connection with teletext.
4. For the five years, 1971–5, annual average expenditure was £620m (1970 prices), and for 1976–80, it was £619m. Estimated from *Advertising*, Summer, 1970 and 1981.
5. The share of advertising expenditure going to directories and yellow pages was only 3.2 per cent in 1980, but this has grown from almost nothing in twenty years (op. cit.). But American experience suggests this form of advertising will become more important, particularly with the growth of videotex (see Brigish, 1981).

   The volume growth of direct mail items moving through the Post Office has increased from 540m items in 1975/6, to 985m in 1980/1 (Lind, 1981).
6. The implications of Prestel and videotex for marketing and retailing, and a review of Prestel's innovation and marketing strategy, have been given in papers belonging to Strathclyde University's Marketing Working Paper Series (Witcher, 1981a, b, c).
7. The French government has adopted a policy of subsidising the

distribution of electronic telephone directories. The Canadian government has also played an active role in subsidising aspects of its videotex development programme. In Britain there has been no direct government subsidy. Telecom have to date spent just over £20m – a small amount when put beside such government subsidies as £80m spent on establishing a sports car company in Northern Ireland to employ 2500 people. In contrast, Videotex is potentially a very large industry indeed.

8. Costs of Prestel receivers vary, but range from £200, for a basic monochrome terminal (without TV), to £1,000 for a large colour TV with teletext. Adaptors are available for about £200. Rental tariffs range from £18 to over £30 a month.

   Telecom charges a connection rate of £15 to an existing telephone line, with a quarterly rental of 15 pence! If a new exchange line is required the charge for installation is £80, with a quarterly rental of £21 for business users. In addition, there is a Prestel levy of £12 per quarter.

   There are three cost elements in information use. There is the normal telephone charge, a time-based charge for using the Prestel computer (4 pence per minute for peak times, a penny a minute for off-peak), and a charge varying from 0 pence to 50 pence per page viewed. There is also the cost of a TV licence, if the user has not already got one.

   It is clear that the price premium for Prestel TV sets are equivalent to the purchase price of most ordinary TVs. Teletext sets, though, are generally about £100 above normal sets, and, of course, once you have the set there is no charge for seeing the information or using computers and telephones.

9. For an account of Sealink's participation in Prestel, see Woodcock (1980).

10. The story and reasons for Viewtel 202's involvement is in Montague (1981).

References

Brigish, A. P. (1981), 'Electronic Yellow Pages', paper presented at *Viewdata 81*, London, Proceedings, Online Conferences.

Fedida, S. and Malik, R. (1979), *The Viewdata Revolution*, Associated Business Press.

Lind, H. (1981), 'Direct mail: at last we have accurate statistics', *Advertising*, 68 (Summer), pp. 26–7.

Montague, P. (1981), 'The electronic newspaper', in R. Winsbury (ed.), *Viewdata in Action*, London, McGraw-Hill (UK), pp. 107–13.

Nicholson, R. and Consterdine, G. (1980), *The Prestel Business*, London, Northwood Books.

Telecom (1980), *Prestel 1980*, London.

Viewdata and TV User (1981), 'News for users', April, p. 5.

Witcher, B. J. (1981a), 'Marketing strategy and Prestel: public service videotex innovation', *Marketing Working Paper*, University of Strathclyde.

— (1981b), 'Public service videotex as a medium for marketing', *Marketing Working Paper*, University of Strathclyde.

— (1981c), 'Tele-information propinquity implications for business trading systems', *Marketing Working Paper*, University of Strathclyde.

Woodcock, D. (1980), 'Sealink, the great Prestel success story', in *Prestel Business Directory*, Financial Times Publishing, October.

6. Microforms in publishing: applications and future involvement with other technologies

PETER ASHBY
Director, Oxford Microform Publications
Oxford, UK

It is certain that microforms have an important part to play in the information technology of the future. Current trends in the application of microfilm-based systems to the transfer of information indicate that links already forged between microforms and other technologies will multiply and strengthen in the next few decades. But before enumerating the ways in which such links are and can be created, some considerations of the nature of the interplay of various information technologies and the role of micrographics therein might prove useful.

To define specific areas within the broad categories of the electronic (rather than the traditional and more tangible) media would be a mistake as barriers between one system or product and another become less distinct with the passing of each information technology conference, associated exhibition, periodical issue, newsletter[1] or any of the numerous trade publicity pieces. Indeed, to write from the standpoint of micrographics may be to fall into this very trap. All indications are that traditional lines of demarcation, if adhered to, provide impediments to the realisation of the broader reality of the information revolution: a revolution whose rapid development may be said to owe more to commercially induced cross-fertilisation than to a natural process of evolution. There is plenty of evidence for a readjustment of interests in professional groups, and not least in the Microform Association of Great Britain, indicating that areas of knowledge are being diversified and that the proposed extension of the all-embracing Graphic Communications Society into an Institute may not be far off. However, despite their essentially interconnected nature, some elements of the new information media have received far more attention and publicity than others, of which microforms must be amongst the more neglected in recent years.

If the test of exposure in the popular mass media is applied to micrographics, then it must compare unfavourably with the other techniques discussed in this book. Certainly, new applications and methods incorporating microforms as reviewed in current awareness periodicals or demonstrated at exhibitions and seminars, may seem at first sight to bear out the notion that 'microform is a child who will never come to maturity'.[2]

There have not been, and are unlikely to be, quantum leaps in the total consumption of reels of microfilm or sheets of microfiche, in the associated chemicals and processing services or in the all-important machinery of creation, dissemination, storage, retrieval, display and printout of the images themselves. However, the steady growth of existing market areas for products and services and the less obvious use of micrographics as an integral part of larger systems will ensure a continuing and substantial role for those concerned with microforms, especially the micropublisher. The pot of gold at the end of the rainbow – a vast consumer market for microforms – is unlikely suddenly to materialise, or at least this is the opinion of those who have looked closely at the market and waited patiently for a revolution (for few, if any, have tried to make it happen!). Moreover, this is not to say that the sum of all specialist microform publishing from archaeology to zoology will not add up to very substantial business for the publishers and purveyors of hardware and software.

There are several reasons for placing such a realistic perspective at the beginning of this chapter. It seems necessary firstly to understand the problems which beset the micropublishing business and larger micrographics industry before thoroughly appreciating its strengths and its enormous, yet to some entirely hidden, potential for the future.

If growth in microform technology is not spectacular and innovation not remarkably novel – remembering that micrographics are over a century old – then it can only rarely be newsworthy, but this does not prevent news and discussion existing, albeit away from the general public's eye. The applications reported and concepts put forward here have been gathered from a wide array of sources, many of them noted as references, while the steady growth in the use of microforms is charted by market research statistics.

Interchange of ideas does not and must not exist merely between micrographics experts. If the elusive 'full maturity' of microforms is to be arrived at, it must be through the co-operation of producer and user, publisher and author, librarian and reader. To this end, groups of producers and users do hold a growing number of meetings under the auspices of various associations[3] and many newsletters, magazines and journals cover the subject. It is an encouraging sign that many of these open discussion articles are not restricted to the specialist microfilm journals but can be found in the press of the wider information sciences and, particularly, the business systems journals.

It must be stressed, therefore, that the majority of those responsible for producing micrographics hardware and services find most of their customers outside the professional information areas (and the probable readership of this volume) and in the rather specialised world of business systems rather than of publishers, librarians and information scientists. The prevailing commercial market demand tends unfortunately to result, on the manufacturers' part, in an over-emphasis on the product itself and a lack of interest in what the medium is being used for and how it should be adapted to suit a wide range of users' requirements. So, a whole area of development is stunted and a yawning gulf exists between the purveyors of microform systems and supplies and a potentially highly significant group of their users.

Take, for example, the case of the historian of medieval art and literature, expert in the iconography of illuminated manuscripts of ninth-century Ireland, who, quite understandably, bypasses the most suitable means of acquiring essential information because it is the novelty of the colour microfiche and associated equipment which is immediately presented to him at a conference exhibition, rather than the work of Macregol, Abbot of Birr.[4] This is an extreme case, but similar instances could be cited at high levels in almost any subject. The common ground, of course, should be one of combined subject interest and economic awareness and these are the areas on which should be focused the combined attention of producers, information scientists and professional users in order to find the optimum application of microform and other technologies to specific problems of information transfer.

Careful plans must be laid for the future at a time when the previously accepted (and, perhaps, still most acceptable) means of conveying information in its final form – the printed journal or monograph – are threatened with economic extinction due to chronic lack of resources both in terms of direct costs (paper, printing and binding) and of indirect expenses (storage and transport). The new (and some not so new) alternatives to the printed book will only be adopted by the discerning user if their constituent parts are properly assembled and presented by the innovating publisher. In short, the new technologies must be as flexible and versatile as the book, and their producers must be aware of the highly sophisticated and diverse needs of their users. If the expectations of those whose interests and deductions had led them to predict a microform explosion have failed to materialise in the past decade, it is because this two-way traffic of information and opinion on microform techniques and requirements has not been properly established amongst all concerned.

In the field of microforms, such significant questions as the speed and method of input and transfer, the preparation of information for the processes which follow and, above all, the suitability of the material in the first place, have not always been thoroughly explored in order that the consumer, who must, after all, both use and pay for the end product, may form a balanced view. Quality of images, technical standards,[5] limitations and potential, and the use of associated equipment (both essential and optional) also need to be put into proper perspective. As these essential factors become more widely understood and appreciated, then it must be agreed that microforms seem at last to have reached the brink of maturity.

The time for full realisation of the potential of microforms, through their integration with other technologies, has arrived and it may, and probably should, all happen without the awareness of the public at large.

The examples which follow have been chosen to illustrate dominant trends in the micrographics and related industries, but also point (in the author's view) to what may result from all concerned taking a look at the widest possible panorama instead of simply staring straight ahead.

The creation of microforms

This low technology process has been operating on a wide scale since the 1930s in large and small organisations by those equipped with varying degrees of skill and with the help of good, indifferent, or, in all too many cases, poor facilities. The types of information-carrying material subjected to microfilming have by the nature of economic demand tended towards time-worn historical documents and poor quality copies of scientific and technical papers. The inevitable opinion of microforms held by the casual observer is consequently disappointed and disappointing and very far from the truth.

The scientific basis of the manufacture of microforms is well documented from the photographic aspect,[6] through the practical processes[7] to the maintenance of all important production standards.[8] Advances continue in a very competitive market where developments from other areas of photographic or business systems technologies are frequently applied.

The growth in awareness of the need for knowledge and training in micrographics and its related areas is reflected in the number of courses run by professional organisations (listed below), by the incorporation of micrographics in formal courses of study in printing, publishing, librarianship and information science, and by the availability of textbooks on the subject.[9,10] The basic principles of micrographics underpin all the examples given below and even if the micrographics element is only small and perhaps invisible or inconsequential to the user, it is nevertheless essential to understand the developments underlying current trends which may, in some cases, represent the cornerstones of greater things to come.

Poor quality recording and lack of standardisation have to date together done more to hamper advances in micrographic applications than the noticeable lack of innovation in equipment and microfilm itself. This situation no longer prevails. A search through the American Standards (described below)[11] reveals that there are seventy-five documents prescribing every aspect of microfilming and micropublishing from preparation of copy through the screen shape and size of viewers to the advertising of micropublications. Papers on workstations from the standpoint of the human reader[12] are as relevant to computer terminal VDUs as they are to microform viewers and this increased awareness of the user's requirements must result in better quality productions.

Despite the welcome introduction of standards, there still exists a wide range of accepted formats for microfiches, both in the arrangement of the images and in their magnification/reduction, ranging from 18–24X through 42–48X to 70–96X, while a fiche may contain as many as 4000 pages. The latter format (PCMI system of the NCR Company) was at one time popular, but criticisms of its 'using a sledgehammer to crack a nut' led to its virtual extinction until a recent revival based on a high technology laboratory[13] and this format is now expected to be confined to the closed systems of industry and commerce where the operator dictates input and retrieval to all users. In the wider world of free information transfer, which must be of most immediate concern, we seem to have enough trouble standardising our electronic media! Encouragingly, microforms are now universally accepted in libraries and research institutions, using viewers and reader-printers catering for 24–28X and 42–48X reduction. Any slight incompatibility between microforms and equipment can usually be accepted with the lens, light-source and screen system, which is very accommodating, unlike the mutually exclusive aspects of much software and hardware found in other technologies.

Having dealt with some potentially worrying aspects of microforms, the vast subject of developments and film types must be approached briefly. Rarely can one innovation be seen in isolation. In mentioning updateable film[14] onto which images can be added, superimposed or removed, consideration has to be given also to the questions of security (and uncopyable film),[15] copyright and archival permanence. Technological advances in film emulsions are aimed at decreasing costs (especially reliance on silver), improving resolution, tonal range and permanence. High resolution colour film hitherto reserved for military mapping is now becoming generally available and opening the way for consumer markets to be developed, e.g. nautical charts for sailing enthusiasts which could be used onboard in an economical and convenient conversion from the large number of bulky maps currently carried by many seafaring yachts.

Microform images from other media

The millions of microform images currently created from existing printed pages or sheets of information are bound to be surpassed by information transferred from other media and this rate of change is accelerating. Computer output onto microform (COM) at 120,000 characters per second is by no means new; it is the growing range of applications, scales of operation and associated variations on the computer microform interface which are worthy of detailed consideration. It is here that other technologies reveal a significant advance over micrographics; however, the innovator may arrive at our area of interest from any of many ill-defined categories – computing, word processing, telecommunications, phototypesetting and so on.

The spectrum of COM-related operations covers a wide range of cost (and, it must be said, quality). At the lower end of the scale is a recently announced unit linked to a word processor which, if it lives up to expectations, will bring microform production within reach of the many involved with personal computers. At the other extreme, and of great significance for publishers and communicators seeking quality and versatility (but still at a remarkably low price when compared with printing), are the full graphic,[16] specialist typographic,[17] and colour capabilities of the latest generation of COM devices.

Advances in computer input microfilm (CIM) based on character recognition now make it possible to reach very high speeds of input from a variety of documents (e.g. catalogue cards) via microfilm. The larger 'on-line' database publishers who feel able to offer full-text retrieval rather than abstracts and bibliographic entries may well turn to microfilm as an intermediary medium, and those who cannot justify the input expense are already using microform for document delivery (see below). The growth areas to watch from the micrographic viewpoint will first become evident here through the advances of other technologies. The speed, accuracy and costs achieved by character recognition-based systems,[18] as used in Aqualine,[19] will combine with other developments and bring microforms into play at various levels. Given that about 75 per cent of backfiles of learned journals are already in microform, it will be interesting to see to what level titles and abstracts are keyboarded for on-line retrieval and how much full text is incorporated and how. The scientific publishers Adonis,[20] for example, have yet to commit to a backfile policy and CIM may wait in the sidelines in vain if other data entry methods are successfully developed (see 'Microforms amongst other media' section below).

One facet of the constant debate in information transfer circles can be highlighted here: what may be technically feasible may never be justified in economic terms. Micropublishers (and those using available services to become micropublishers) therefore view the volatile cost margin with interest. The literature of law may be another indicator: will LEXIS, EUROLEX and the like extend back to medieval case law? Are there grounds for providing a more sophisticated system than sets of microfiches with or without printed guides (see below)?

Microform to hardcopy and back again

The print-out of microimages is a natural extension of the micropublishing system, but not, as is sometimes the case, without proper awareness of its limitations. The price of creating a good quality paper enlargement 'on demand' is usually so low in comparison to alternatives that the quality, cost and therefore availability of plain paper reader-printers is improving in response. Where a paper original never existed, an ever more frequent occurrence, micropublishing comes into its own, removing the expensive printer from the scene altogether and yet proving capable of things which electronic publishing cannot achieve either technically or economically.

The information used to create visual records from laser plotters can be transferred on-line and recreated remotely via CIM and other devices, but, the question inevitably arises, at what cost and quality? Considerable advances in facsimile transmission of microform images have been made in Japan and a recent example from Australia of computer-generated fiche within a printed publication, with the added dimension of colour (an increasingly essential factor), has achieved a satisfactory compromise between cost and quality, providing an exciting indicator of things to come. The example in question, an atlas of 540 large-scale maps depicting details of complex land use change in the state of Victoria,[21] has been published in combined text and fiche form. Census data and environmental factors are computerised, seven colour codes added, and the resultant information manipulated against specially devised programmes and run through a COMP80/2 computer output microfilmer to create a series of maps on monochrome fiches and on 35mm frames which, when superimposed and re-formatted onto colour film by a step-and-repeat method, produce colour fiches presenting information with colour coded and spatial dimensions. It is expected that this innovative process will be written up in detail and that many more applications will follow.

The high quality and large number of colour illustrations required for the atlas project outlined above and for many other examples of low print run publications must rule out the economic viability of a printed version and open the way for microforms. Scientific, technical and medical publications incorporating colour fiches generated through electronic data processing, microscopy or conventional photography will benefit from new colour microfiche production methods and better viewers and reader-printers. In the area of the fine arts, micropublishers are less likely to face serious competition from electronic methods and, particularly, from video discs because of the problems inherent in these types of technology both in resolution and in length of viewing time for a single image (enlarged far more flexibly by projection than by CRT). As for working copies printed from the microforms, established colour processing laboratories can enlarge from camera originals for best results, but paper printouts may also prove an area to watch out for in the future if the colour-photocopier and polaroid-principle instant slide copier incorporate microform images as input.

Of great potential relevance to all lines of argument developed so far is the advent of the video still camera. No larger than a single lens reflex amateur photographer's camera, and very similar to it in appearance, this Japanese produced unit records any visual image electronically – and presumably the initial information storage system in the camera will be compatible with other electronic data systems. The understandably sceptical view of this development put forward by the Eastmann Kodak

Company[22] is edifying, but it does seem, nevertheless, that this is another area of potential involvement with electronics about which micrographics users should keep themselves informed.

Microforms amongst other media

The ultimate convenience of the individual recipients may dictate the extent to which images are transferred from hard copy to digital form to microform and back again. If large geographical distances need to be covered and time is critical, then electronic transmission or the sending of microforms by airmail will be preferable to the conveyance by air parcel of volumes of paper. The interchangeability of ink on paper, images on microform or electronically stored information is one of the most exciting prospects to come out of the compatibility of information interfaces.

The value of microforms for the future must lie in the range of options and combinations offered in any given case, and this essential flexibility hinges on the twin factors of the economies and convenience of storage and retrieval (expanded in the following section). Before considering these fundamentals, however, and at the risk of overlapping with other contributors to this volume, a brief survey of some seemingly independent developments in information handling and their impact on, or interface with, microforms is essential.

The tortuous journey of information from authors through editors, database operators or publishers, and then through distributors and information stores to the ultimate users is blessed by too many innovations to catalogue here and only an on-line system could cope with all the new products and services available. Considered here are merely some recent and some more established additions to the communications network possessing a particular relevance to micrographics.

Comprehensive text processing (CTP) claims to be able to handle the vast wealth of human knowledge existing in printed form – and by implication its microform version – ingesting from 6- to 24-point typefaces of all styles, including cyrillic, a 25 per cent cost saving on re-keyboarding. CTP is claiming to be four times as fast as traditional input and 95 per cent error free, with built-in routines for proof-reading of style and spelling. It is believed that such input services are only available in one facility in Europe at present.

The basic equipment required for CTP, the Kurzweil Reading Machine, is also being developed as 'Omnifont',[23] a blind person's 'display' unit giving voice to the text displayed. No doubt at some point this process will be reversed to create digitised text from voice input (indeed, experiments are currently being carried out in Japan on 'voicewriters'). Another device sidestepping the conventional typewriter and its 'QWERTY' keyboard arrangement is the Microwriter.[24] The Microwriter works through a grid of six buttons on a hand-held unit facilitating alphanumeric input in electronic form which, alongside CIM and CTP, will bring universal information transfer to the home. With the aid of a Husky 144 portable computer,[25] a telephone and a portable microfiche reader, any individual in Great Britain will be able to achieve total involvement with all information transfer methods, while remaining mobile.

The question of why a low-technology medium such as microform should remain integrated in such a sophisticated scenario as that outlined above is a valid one. There is no question, however, that the printed book or periodical will retain its place and the answer must be that each medium has its own place and role within the total system.

As the economics of microform production, storage and dissemination become evident to all concerned with the process of information transfer from author to reader, so poor and inappropriate uses of micrographics will become fewer, and well planned, successful applications commonplace. The process of information transfer is under constant debate and the proper role of microforms is continually being reassessed, but this is a healthy sign.

Microform benefits

Beyond the well-worn and often less than relevant space-saving argument, the justifications for incorporating microforms with other methods of information transfer are more interesting. Of course the storage of books and journals in libraries carries very significant costs and these are seen to be reduced in a very tangible way when text-fiche publications are acquired.[26]

Advances in microform storage and filing systems serve to increase economies and improve presentation and use.

Less tangible advantages, usually considered only by the micropublishers, hinge on the economics of short-run or on-demand publication and, increasingly, on the value of information origination directly onto microforms. Security has many meanings in terms of information, and microform has long been associated with the archival copying of documents. Also microforms are less easily browsed through than paper copies by unauthorised personnel. In addition to these security features, microforms compare very favourably with electronic systems in terms of cost of storage and of speed (and cost) of retrieval and searching.

DatagraphiX Inc., one of the major US manufacturers of COM devices, have undertaken a study of competing technology-based storage and transfer systems, comparing costs, capacity, access time in milliseconds, susceptibility to information change, storage time in years and the user's need to read only or to read and write (see Table 1). Even with the spectacular improvements anticipated with the introduction of the semi-conductor/chip storage methods, the cost of electronic storage remains several orders of magnitude greater than that of microform.[27]

Outside the obvious areas of business and government systems, those responsible for storing archival data for subsequent retrieval and use may feel happiest with the miniaturised page format which is made eye-readable with nothing more complex than a system of lenses and is not in danger of being wiped out. Indeed, total reliance on electronics may be risky in certain regions and organisations. Much COM throughput is now acknowledged to be primarily for security purposes.

It will be interesting to compare the relative cost curves created from the transfer of large databases via satellite or via microform with due regard to speed and security.

Full-text searching of microform is of course the deciding factor in many cases. The next logical topic for consideration is therefore the retrieval of information from microforms, whether they stand alone or form an integral part of an information system.

Table 1. Competitive storage technologies. Hierarchy of storage needs shows significant cost differences for the different performance characteristics of technologies shown.

A	B	C	D	E	F	G	H	I	J
	Cost per megabyte				Storage Base	Access Time	Volatile or	Storage Life	Read–Write or
Technology	1980	1983	1985	1990	(megabytes)	(seconds)	Non-Volatile	(years)	Read Only
Semiconductor	$15,000	$7,500	$3,750	$2,500	1	1×10^{-7}	V	0	R–W
Ebam	NA	NA	12,500	12,500	8	1×10^{-5}	NV		R–W
CCD	16,000	4,300	2,400	800	1	1×10^{-4}	V	0	R–W
Bubble	10,000	2,900	1,250	400	1	1×10^{-3}	NV	1	R–W
Magnetic Disk	41	20	10	5	570	2×10^{-2}	NV	1	R–W
Magnetic Tape	4.21	3.61	3.25	2.52	5,000	7×10^{-1}	NV	1	R–W
Magnetic Mass Store	5.60	4.80	4.33	3.35	462,000	1.6×10^{1}	NV	1	R–W
Optical Disk	NA	8.00	8.00	8.00	2,500	3×10^{-1}	NV	10	RO
Optical Disk Pack	NA	NA	1.60	1.60	125,000	7.5×10^{-2}	NV	10	RO
Optical Disk Mass Store	NA	NA	NA	0.008	25,000,000	3×10^{0}	NV	10	RO
COM 48X	0.67	0.65	0.64	0.61	250	1×10^{1}	NV	100	RO
COM 48X	0.09	0.09	0.08	0.08	2,500	1×10^{1}	NV	100	RO
COM 96X	NA	NA	0.23	0.23	1,000	1×10^{1}	NV	100	RO
COM 96X	NA	NA	0.03	0.03	10,000	1×10^{1}	NV	100	RO

Searching and retrieval from microforms

The greater the editorial and production investment made by the microform publisher, the higher the unit cost of the result. The more effort which is put into ordering and indexing the data committed to microform, the easier it is to search and retrieve desired information. At the lowest level, this indexing simply takes the form of a printed contents sheet accompanying the fiches or films. The trend of combining a fuller text with fiches in a bound book is frequently praised but is dogged by misconceptions about the supposedly inexpensive nature of such a micropublication, which, in fact, requires intensive editorial preparation, plus the expense of the conventionally printed element, all in addition to production costs for the microfiches themselves. 'Value' is all too often not related to information content and its level of sophistication in terms of accessibility and updateability.

A recently published scientific handbook on the C4 Hydrocarbons contains cross-references from an index and extended contents in bound book form to c. 1200 microformed pages.[28] This publication was produced manually, but word processors will shortly be used to cut down the editorial investment.

Moving beyond text–fiche publications which require only a simple microform viewer for their interpretation, the rapidly developing area of on-line searching of catalogue data must be reviewed. Only the briefest mention of a few examples is possible here, but it is as an integral and often invisible part of most researchers use of reference tools that microforms are likely to achieve their greatest and most innovative development.

Where extant documents or detailed visual records with line and, particularly, tone and colour illustrations are concerned, a microfiche store may be most appropriate. Readily available explanations of document organisation and filing, suited especially to the needs of business systems, are already on the market, and the Kodak IMT[29] equipment (for Computer Aided Retrieval of microform) provides the basis for encoding key descriptors of each document or page as it is microfilmed. If no pre-sorting is to be undertaken, then a RAM (Random Access Memory) capability in the associated computer hard- and software is essential. Closed circuit television and on-line searching can vastly extend access to a single location information bank and a French producer has demonstrated this already in commercial applications.[30] The trend will doubtless be towards library and open user applications as and when such markets can be identified and properly served by publishers and, most important, can afford to pay the price of constantly updated information in this form.

The most successful commercially operated electronic information services are likely to be sufficiently secure financially to pioneer techniques which can later be applied, in refined form, to more specialised and less well funded areas of research. Textline[31] provides economic and business information to the financial community and fully indexed abstracts are keyboarded daily, adding to a database which, as it broadens, deepens and covers a greater timespan, becomes increasingly valuable. The cost of abstracting and entering (albeit a means of sidestepping copyright questions) is, however, expensive in respect of demand for a large number of items which must inevitably be seldom or never used.

The day cannot be far off, therefore, when an economic decision will be arrived at by the risk-taking publisher not to include certain marginal information either in full text or abstracted form. Perhaps the answer is to support bibliographic entries with abstracts and full text in microform which can be distributed automatically to subscribers a system successfully implemented by the Soviet and East European Economics Abstracting Service, ABSEES. Some users may only commit to subject profiles while others could rely on a document delivery system only.

A fine example of the use of microforms in the transfer of information was explored in the course of research for this review. The only available printed listing of standards applicable to micrographics was known to be out of date and a new catalogue had to be acquired by post from a Maryland US address.[33] Once it was known that this publisher's catalogue was constantly updated in machine-readable form on the Lockheed computer in Palo Alto, California and could be interrogated via the Dialog on-line search facility in Great Britain, a ten-second search was made. This search revealed the existence of seventy-five relevant standards, and all these catalogue entries were printed out on ten feet of paper in three minutes at a cost of less than £10. It is at this point that microforms enter the system. Once in possession of a definitive and up-to-date list of titles, individual documents can be ordered by using an electronic maildrop facility. An order placed by 3 p.m. Eastern US time can be fulfilled on the same day and the full item sent by airmail in microform. The three- to four-day delay and the low cost of diazo fiches and airmail postage have proved perfectly acceptable to both user and publisher in this case. A larger database, like the NTIS[34] (National Technical Information Service), with 800,000 abstracts, experiences, according to its UK agent, much less demand for microfiches, and full documents on paper are preferred, but this may well be a function of limited customer awareness. Records of UK businesses from Companies House[35] are supplied in microform at remarkably low cost and provide an excellent means of microform education amongst financial and legal professionals and business people as no choice of medium is offered. Certainly, in terms of expenditure, the argument for document delivery in microform is a powerful one, as price and time differentials both show considerable savings on paper costs.

There may be systems where microforms are suitable but the three- to four-day delay for trans-Atlantic airmail transfer proves unacceptable, and, in cases such as these, the BNF/Image Systems On-Line Fiche Retrieval Unit[36] may provide an example of an inexpensive compromise. The BNF retrieval unit consists of a carousel of microfiches, totalling up to 187,200 pages, which can be constantly updated by post and can be accessed by remote control by searching a descriptive catalogue on-line. Having identified an entry on the CRT, a search can be operated on a sealed fiche viewer within which the appropriate fiche is selected by CAR and the precise grid reference located and displayed for screen viewing and print-out. (The on-line bibliographic entry would of course indicate the availability of a new microfiche yet to be filed.) The happy compromise reached in the BNF system should provide an indicator of things to come, especially in the field of non-textual information.

In some areas of research, funding may be sufficient to warrant alphanumeric and digital display on line. The Video Patsearch service of Pergamon's Infoline system,[37] for example, has a dual CRT on one search console with print-out, while Derwent's competitive system[38] uses microfiches and 16mm microfilm in support of on-line. It will be interesting to see if all visual material is computerised or if microfiches will provide some if not all of the information, in the many cases of documents likely to be demanded.

Nor is the microform/computer interface confined to the highest levels of commerce and research. Microfiches are used in training and teaching in conjunction with an audio tape, a combination pioneered by Revox.[39] Sequences of illustrations from microform images are controlled by a keyboard with commentary/music and programmed instructions. In most of these systems the physical form of the microfiche is as irrelevant as the tape cassette in an audio or video system, the user simply interacts with the information presented to him and, having loaded the machine with the software, forgets it. In the case of the above mentioned retrieval system, where a carousel of microfiches is searched remotely, the user may never come into contact with the medium of transfer at all, as periodic updates would be loaded by the information specialist or technician. In one newspaper cuttings service, based on remote searching and retrieval of microimages, the search and printout function is carried out by the library staff, and journalists in receipt of a sheaf of papers within a few minutes of making a request may not be aware of any of the technology employed in its fulfilment.

Display and use of microforms

Despite the cases cited above, most researchers will be directly involved with the medium and need to use microform viewers and reader-printers alongside modems, keyboards and CRTs. Not enough is being done on the microform side to improve the ergonomics, design and cost of units. NATO is known to be interested in the improvement of the personal microfiche viewer and it is to this type of user body that the industry must look for the development of wider and more demanding markets. In the present unsatisfactory situation, business systems applications seem to keep manufacturers happy with an easily defined and satisfied market.

Matters have reached a sorry state when readers of such

specialised texts as *The Dissected Horse*,[40] which incorporates ten colour microfiches of gross pathological illustrations, are shown how to construct a peep viewer out of a cardboard holder and a cheap lens! The 1980s must see improvements in this approach to microform viewing outside libraries. The Japanese Fuji viewer has led the field in portable viewers for the last five years and the Polaprint device already described holds promise for private print-out, if developed.

Micrographics will of course continue to be in close competition with photocopying and short-run printing techniques and especially the new and economical laser/link jet printing and 'desk-top' perfect binding methods. However, with improvements in presentation and use and the increasing involvement of microforms with other technologies, their continued growth should be assured.

Indications of the future involvement of micrographics in information transfer

High technology based systems will probably be found first in specialised fields, some of which will be limited both economically and practically to the frontiers of research. Few manufacturers of equipment, publishers or operators of systems would not wish for larger markets and these cannot exist in the specialist areas at the very highest levels of each subject. Only wider markets can fund further developments and improvements in products. The home computer has arrived in high street stores and software publishers now sell discs and cassettes by mail order, but where are the microforms?

Many larger booksellers,[41,42] banks, building societies, garages and libraries have on-line and off-line computer terminals and many more smaller branches of such organisations have microfiche readers. Other work places are becoming involved also as relevant information is conveniently transferred to microform. In North America, estate agents[43] and travel agents[44] use microfiches to display a far greater range of illustrations than brochures allow, even if booking data is dealt with on line – and videocassettes displaying all aspects of a holiday resort are becoming commonplace. In the vast field of advertising, 'artfiche'[45] is an economical means of bringing a large range of high quality images by advertising photographers to their clients' attention.

Just as the information technologies are being recognised in printing, publishing and librarianship courses, so they are being seen educationalists in other areas as fundamental in furthering the learning process and preparing for work. Students will soon become familiar at middle school level with the interrogating of databases, the searching of on-line catalogues and with the principles of document retrieval.

A broad educational approach to computers and micrographics is being taken by more than one UK polytechnic, and a scheme based on the popular Tandy home computer offers individual access to foreign press cuttings, never normally accessible in Great Britain, through a reference and abstracting service on microfiches.

But perhaps it is in the field of the fine arts[46] that microforms have the most ground to make up, in order to assure a very promising future for themselves. Whereas press cuttings may be transferred entirely to electronic form in years to come, microfiches must become increasingly important for visual images, an area where their superior resolution will never be replaced by a videocassette file or videodisc.[47]

Opportunities and challenges

It is appropriate to conclude on the point of 'still' image quality, which however inexpensive and convenient it may be to create, must nevertheless be transferred to the user in such a way that the quality of the original is unimpaired. The printing industry has refined its techniques over five hundred years and, by as many as ten passes through the press, can produce a facsimile of a work of art capable of satisfying the most demanding expert or art lover. The same exacting demands can be met, although perhaps in less lasting form, by the relatively new photographic process, which has managed to achieve its results in less than two centuries of accelerated progress. The big question remains, however, as to whether electronically based technologies can ever equal the printer's skill or the fidelity of the micropublisher's continuous tone reproduction.

At the beginning of this chapter it was noted that the viewing of other technologies from the standpoint of microforms could prove deceptive and that many of the other information handling techniques could equally well be put forward as being mainstream in the current trends outlined here. The ultimate test is, of course, that of time, and a brief glance at the dates of the notes at the end of this chapter will serve to indicate the infancy of our subject.

The challenge of the last two decades of the twentieth century is to apply what technology has provided and improve on techniques in a constantly changing and rapidly growing market for information. Full integration of each of the so-called separate technologies referred to here will soon see the redundancy of such a volume as this. 1982, the Year of Information Technology in the UK, should, as an exercise in communication itself, add to the efforts of many individuals who, like the scholar of ninth century manuscripts introduced at the beginning of this chapter, would welcome the day when the message and not the medium was of all-consuming interest.

Notes

1 Some periodicals which cover micrographics as a means of information storage and transfer:

British Journal of Educational Technology (London, Council for Educational Technology, 1975-)

Communication Technology Impact (Oxford, Elsevier, 1978-)

Computer Languages (Oxford, Pergamon, 1976-)

Le Courier de la Microcopie (Paris, Micro-Journal, 1974-)

Euronet Diane News (Luxembourg, DGXIII EEC)

Information Processing and Management (Oxford, Pergamon, 1963-)

Information Services and Use (Amsterdam, North Holland, 1980-)

Information Systems (Oxford, Pergamon, 1975-)

Information Technology Research and Development (Sevenoaks, Kent, UK, Butterworth, 1982-)

International Journal of Micrographics and Video Technology (Oxford, Pergamon, 1982-)

International Micrographics Congress Journal (Bethseda, MD, USA, I.M.C., 1971-)

Journal of Information Science (Amsterdam, North Holland, 1978-)

Library Acquisitions, Practice and Theory (Oxford, Pergamon, 1977-)

Microdoc (Guildford, Surrey UK, Microfilm Association of Great Britain, 1962–). Now *I. J. M. & V. T.*, listed above)
Microform Review (Westpoint, CT, USA, Microform Review, 1972–)
Micrographs Newsletter (Wykakil, New York, Microfilm Publishers Inc., 1969–)
Microinfo (Alton, Hants, UK, Microinfo, 1970–)
Micro Notes (Ottawa, The Canadian Micrographic Society, 1972–)
Micropublishing of Current Periodicals (Toronto, Canada, University of Toronto Press, 1968–)
On-Line Review (Oxford, Learned Information, 1976–)
Reprographics Quarterly (formerly NRCd Bulletin) (Bayfordbury, Herts, UK, National Reprographics Centre for Documentation, 1967–)
Scholarly Publishing (Toronto, Canada, University of Toronto Press, 1968–)
Social Science Information Studies (Sevenoaks, Kent UK, Butterworth, 1980–)
STM Innovation Bulletin (Amsterdam, Scientific, Technical & Medical Publishers, 1975–)
Videoinfo (Alton, Hants, UK, Microinfo, 1981–)
Visual Resources (Oxford, Oxford Microform Publications, 1980–)

2. E. Gray, 'Electronic systems and the future of paper and microform publishing', *Micropublishing of Current Periodicals*, 4, (3), 1980.
3. Some organisations which hold seminars etc. covering micrographics:
ASLIB, 3 Belgrave Sq., London SW1X 8PL.
Association of Database Producers, 30 Austenwood Close, Chalfont St Peter, Bucks, UK.
British Computer Society, 13 Mansfield Street, London W1.
Business Equipment Trade Association, 8 Southampton Place, London WC1A 2EF.
European Association of Information Services, Rose Cottage, Moulsoe, Bucks, UK.
European Information Providers Association, Rodenstraat 125, B1630, Linkebeek, Belgium.
Federation Internationale de Documentation, 7 Hofweg, 2511 AA, The Hague, Netherlands.
Institute of Information Scientists, 62 London Rd, Reading RG1 5AG, Berks, UK.
Institute of Scientific and Technical Communicators, 17 Bluebridge Ave., Hatfield, Herts, UK.
International Micrographics Congress, P.O. Box 34404, Bethesda, Maryland MD 20817, USA.
Library Association, 7 Ridgmount St., London WC1.

Microform Association of Great Britain, Dellfield, Pednor, Chesham, Bucks, UK.

Microform Review, Saugatuck Station, Westport, CT, USA.

Microinfo, Newman Lane, Alton, Hants, UK.

National Micrographics Association, 8719 Colesville Rd, Silver Spring, MD 20910, USA.

NRCd, Bayfordbury, Hertford SG13 8LD, Herts., UK.

On-Line Conferences, Learned Information, Wooton, Abingdon, Oxford, UK.

4. *The Macregol or Rushworth Gospels (MS. Auct.D.2.19)*, Series I, Major Treasures in the Bodleian Library, no. 10, Oxford, Oxford Microform Publications, 1979.
5. Some sources of technical standards in micrographics:
   American Standards (ANSI), *see* note 11.
   British Standards Institute (BSI), 2 Park Street, London W1A 2BS.
   International Standards Organisation (ISO), ISO/TC, 171, Tour Europe, Cedex 7, Paris.
6. G. W. Stephens, *Microphotography*, London, Chapman & Hall. Now out of print but available from Oxford Microform.
7. M. Gunn, *Document Microphotography*, London, Focal Press, 1982.
8. *BS 5444: 1977 Recommendations for preparation of copy for microcopying*, London, BSI, 1977.
9. S. J. Teague, *Microform Librarianship*, London, Butterworth, 2nd ed., 1979.
10. P. Ashby and R. Campbell, *Microform Publishing*, London, Butterworth, 1979.
11. *DIMS (Direct Index Microfiche System) Standards and Specifications* (44,000 documents and weekly revision service), National Standards Association Inc., 5161, River Road, Bethseda, MD 20816, USA.
12. R. M. Landau, 'Workstations: the human factors approach to office automation', *IMC Journal*, 16, (4), 1980.
13. UMF Systems, 5221 Grosvenor Boulevard, Los Angeles, CA. 90066, USA.
14. D. R. Wolf, 'The technologies and role of updateable micrographics', *Journal of Micrographics*, August, 1981.
15. Bexford Microfilm Products, Manningtree, Essex, UK.
16. Laser Scan IMCP-1 plotter, Imtec Ltd, Nailsen, Bristol, UK.
17. Hell Digiset, Bemrose Information Services, Derby, UK.
18. P.G. Jennings et al., 'Data capture by optical scanning of published material for database enhancement', *Program*, Feb., 1982.
19. L. E. Newman and D. Haynes, *Aqualine Online User Guide*, Water Research Centre, Marlow, Bucks, 1981.
20. Adonis, a project for the electronic delivery of documents. B. T. Stern

Elsevier Science Publishers, P.O. Box 2400, 1000 CK, Amsterdam, Netherlands.

21. J. S. Massey, *A Computer Plotted Microfiche Atlas, Agricultural and Pastoral Land Use, Victoria, Australia*, Department of Geography, University of Melbourne, 1981.
22. 'Video stills: it's hardly Eureka!', *Professional Photographer*, Nov., 1981.
23. Omnifont – the Kurzweil Reading Machine, 12 High St, Chalfont St Giles, Bucks, HP8 4QA, UK.
24. Microwriter, Wandle Way, Mitcham, Surrey, UK.
25. Husky 144, DVW Microelectronics, 10 The Quadrant, Coventry, CV1 2EL, UK.
26. 'Text-Fiche' is the registered name of a list of publications on art historical themes published by University of Chicago Press (Ellis Ave., Chicago, Illinois, USA) but is becoming more widely applied to combined print and microform titles.
27. B. Suiter, 'COM price/performance leads storage media', *Computer World*, Sept., 1980.
28. E. Hancock (ed.), *The C4 Hydrocarbons and their Industrial Derivatives*, Oxford and London, Oxford Microform Publications/Benn, 1980.
29. 'Electronic filing from Kodak', *Kodak Business Systems News*, Jan., 1982.
30. Videocommunication: videodoc, videoplan, videocheque, videoconference, videoreproduction, Cédam, 173, Rue de Crimée, Paris 75019.
31. Textline, Finsbury Data Services Ltd, 68/74 Carter Lane, London EC4V 5EA.
32. *ABSEES*, Oxford Microform Publications, Paradise St, Oxford.
33. See note 11.
34. National Technical Information Service (NTIS). UK agent is Microinfo, P.O. Box 3, Alton, Hants GU34 2PG, UK.
35. Companies House, Crown Way, Cardiff CF4 3UZ.
36. BNF Information Services, BNF Metals Technology Centre, Wantage, Oxfordshire, OX12 9BJ, UK.
37. Video Patsearch, Infoline, Brettenham House, Lancaster Place, London WC2E 7EN.
38. Derwent Publications, 128 Theobalds Road, London WC1X 8RP.
39. Revox, Willi Studer GmbH, CH-8105 Regendorf, Switzerland.
40. W. O. Sack and L. C. Sadler, *The Dissected Horse*, Cornell University, 1976.
41. *British Books in Print on Microfiche*, J. Whitaker & Sons, 12 Dyolt Street, London WC14 1DF.

42. *Books in Print on Microfiche*, Bowker, P.O. Box 5, Epping, Essex CM16 4BU, UK.
43. 'Microfiche gives builders quick access to thousands of custom house plans', *IMC Journal*, 16, 1980.
44. Infox, 45 Sheen Lane, London SW14 8AB.
45. Artfiche, Arthur Wilson, Artfully Photographic, 32-4 Gt Marlborough St, London W1.
46. P. Ashby, 'Illustrated reference teaching and beyond', *Microform Review*, 8, (3), 1979.
47. E. M. Lewis, 'Video scan picture searching', *Visual Resources*, 1, (1), 1980.

7. HIGH QUALITY COMPUTER PRINTING

HEATHER BROWN
Computing Laboratory, University of Kent,
Canterbury, UK

Print it as it stands – beautifully.
Henry James

Introduction

Printing produced by computers is often shoddy. Most people can tell at a glance which parts of their bills or bank statements are printed by a computer – the parts that are badly printed and awkward to read. This may not matter too much for bills, but the current explosion in office word-processing systems is already leading to a corresponding explosion in the number and variety of badly printed documents being produced.

When speculating on the future of the printed word, one of the safest predictions to make is that the current trend towards producing documents by computer will not only continue, but will accelerate. Does this mean that we can look forward to a future where we are deluged with inferior print? The main purpose of this paper is to show that the answer to that question is 'no'.

New systems take several years to emerge from research and development departments into general use. Thus the systems that are going to make an impact in the next five or ten years already exist. And it is here, among the experimental systems, that the emphasis is turning towards high quality printing.

The two worlds of document preparation

Document preparation systems on computers are not new – they have been with us for twenty years. The great variety of systems that exist today have grown up in two different worlds. It is unfortunate that they have done so in such a way that each world tends to despise the other for its weaknesses without learning from its strengths. A quick look at the current situation will help to explain the significance of the new developments.

On one side there is the general computing and office world which produces documents ranging from office memos to long technical manuals and book manuscripts. Most general-purpose computers and word-processors offer document preparation programs which will lay text out neatly on a page, hyphenate words where necessary, split long texts into separate pages, insert page numbers and running titles at the top of each page, and so on. The document preparation system may also offer 'macro' facilities which allow the user to tailor the system to his own needs and to make systematic changes to a long document easily and quickly.

The big disadvantage of these systems is that they are geared to printing the final document on a standard computer printer. Typically these printers are of low quality and, not only is their repertoire of characters very limited, but they are only capable of printing characters in a single size. The resulting documents may be laid out beautifully on the page, but the actual quality and flexibility of the printing is, at best, like that of a standard typewriter. Figure 1 shows an extract from a document printed on a high-speed lineprinter. Figure 2 shows the same extract printed on a higher quality, but much slower, daisy-wheel printer. In both cases the spacing of characters is inflexible.

```
                         I
          'Elementary Rules of Usage

1. Form the possessive singular of nouns by adding 's.

   Follow  this  rule  whatever  the  final  consonant. Thus
write,

        Charles's friend
        Burn's poems
        the witch's malice

Exceptions are the possessives of ancient  proper  names  in
"-es" and "-is", the possessive "Jesus'", and such forms as
"for conscience' sake", "for righteousness' sake". But  such
forms as "Moses' laws", "Isis' temple" are commonly replaced
by
```

Figure 1. Lineprinter output

Every character – even a space – occupies the same area. The only way to vary the spacing between words is to insert extra fixed-size spaces. These inflexible printers are the ones that have given computer printing a bad name.

The other world is the increasingly computerised world of the professional printer and the manufacturer of typesetting equipment. In this world the emphasis is on the flexibility and high quality of printing. Typesetters allow a wide choice of different styles and sizes of type and can cope with fine details of spacing between characters to improve the appearance of the final result. Figure 3 shows a typeset version of Figures 1 and 2, and Figure 4 shows a few examples of the different character fonts which are likely to be available. Unfortunately, printer's systems lack many of the facilities which make it so easy to write and rewrite a document using a document preparation system on a general-purpose computer. Typically there are no facilities for splitting documents into separate pages – the output consists of a continuous strip of text which needs to be cut up into pages and needs page numbers and headings to be inserted by hand. This may be fine for a printer who is used to the process, but the very idea fills the average office user with horror. Also, these systems tend to need a skilled operator and they are not readily available to the average computer user.

Because of the problems created by the two separate worlds an absurd situation has grown up. The author of a technical

```
                         I
            Elementary Rules of Usage


1. Form the possessive singular of nouns by adding 's.

   Follow this rule whatever the final consonant. Thus
write,

          Charles's friend
          Burn's poems
          the witch's malice

Exceptions are the possessives of ancient proper names in
"-es" and "-is", the possessive "Jesus'", and such forms as
"for conscience' sake", "for righteousness' sake". But such
forms as "Moses' laws", "Isis' temple" are commonly replaced
by
```

Figure 2. Daisy-wheel printer output

I

Elementary Rules of Usage

1. Form the possessive singular of nouns by adding 's.

Follow this rule whatever the final consonant. Thus write,

Charles's friend

Burn's poems

the witch's malice

Exceptions are the possessives of ancient proper names in *-es* and *-is*, the possessive *Jesus'*, and such forms as *for conscience' sake*, *for righteousness' sake*. But such forms as *Moses' laws*, *Isis' temple* are commonly replaced by

Figure 3. Properly typeset version of the text shown in Figures 1 and 2

ABCD abcd 1234 %&();? fi fl

ABCD abcd 1234 %&();? fi fl

ABCD abcd 1234 %&();? fi fl

ABCD abcd 1234 %&();? fi fl

ABCD abcd 1234 %&();? fi fl

ABCD abcd 1234 %&();? fi fl

ABCD abcd 1234 %&();? fi fl

ABCD abcd 1234 %&();? fi fl

ABCD abcd 1234 %&();? fi fl

ABCD abcd 1234 %&();? fi fl

ABCD abcd 1234 %&();? fi fl

ABCD abcd 1234 %&();? fi fl

ABCD abcd 1234 %&(); ? fi fl

ΑΒΓΔ αβγδ

абцд АБЦД

Figure 4. Samples of character fonts available on a typesetter. The upper group shows the same character font set in different sizes; the lower group shows several different fonts.

paper or book frequently prepares the manuscript himself (or maybe his secretary does) using his local document preparation system. The manuscript is rewritten and corrected until all the errors have been eliminated and it finally reaches a satisfactory form. Currently there is no way that the output from the local system can be fed directly to a high quality printer or typesetter, so a printed version is sent off to the publisher with some special instructions to overcome the deficiencies of the local printer. The author may, for example, request that all section headings should be converted to bold type and that all underlined words should be turned into italics. The publisher takes the printed manuscript and has it retyped into his own typesetting system. At this stage a whole new set of errors and layout problems will inevitably be introduced and the process of proof reading and correcting will need to be undertaken a second time.

The new technologies

The dramatic fall in the cost of computing power is producing changes in both worlds. New printing devices are available and computing power can be provided to drive them economically.

A new range of comparatively cheap laser printers is just coming on to the market. These printers work on the 'dot-matrix' principle and have a resolution of around 300 dots/inch. They can be used to print any desired character shapes – the individual dots making up the shapes are so close to one another that they are almost indistinguishable. Given the considerable computing power necessary to drive them, these printers can produce good quality print and have the flexibility needed to produce a potentially unlimited range of special characters including mathematical symbols, Greek characters and Japanese characters. Figure 5 shows the same extract as Figures 1, 2 and 3, but this time it is printed on a laser printer with a resolution of 240 dots/inch.

A similar development is the appearance of the high-resolution graphics screens (around 100 dots/inch) which are already beginning to take over from the fuzzy and restricted computer screens that most people are familiar with. The capabilities of these screens allow dramatic improvements in the way

I

Elementary Rules of Usage

1. Form the possessive singular of nouns by adding 's.

Follow this rule whatever the final consonant. Thus write,

Charles's friend
Burns's poems
the witch's malice

Exceptions are the possessives of ancient proper names in *-es* and *-is*, the possessive *Jesus'*, and such forms as *for conscience' sake, for righteousness' sake.* But such forms as *Moses' Laws, Isis' temple* are commonly replaced by

Figure 5. Laser printer output

information can be presented.[1] Many pleasing effects can be produced by displaying separate 'windows' of information simultaneously or by superimposing images. A typical screen layout is shown in Figure 6. One window is probably a control window where the computer displays messages to the user and provides a record of recent actions. The other windows are best thought of as pieces of paper on a desk. They can be moved around, covered by others, brought back on top, and so on. The image in a window need not be static. If really desired, it could represent a digital clock or an animated cartoon. While not directly relevant to printing, these screens are important in document preparation systems. They can show different character fonts and they allow proof copies of documents to be inspected. By displaying several pages simultaneously, a user can pick out errors or inconsistencies of style or content before printing.

Just as advances in technology are producing better printers within computer systems, they are also producing better computers within typesetting systems. The first computer-controlled typesetters reproduced characters by optical methods from master shapes stored on films or discs, but it is now cheaper to use digital methods to reproduce characters as patterns of dots from digitised masters.[2] Typesetters have a resolution of

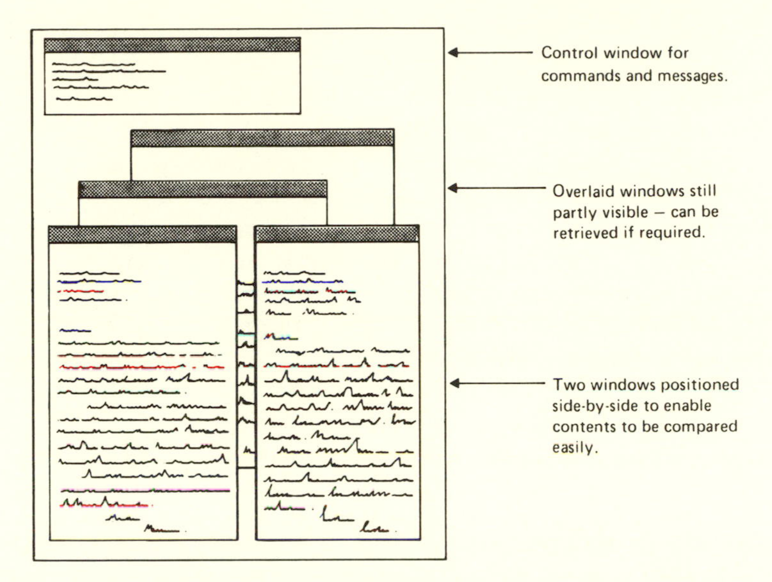

Figure 6. Use of windows on a high-resolution screen. The A4 size screen is split into a number of windows, each with a caption in the top frame.

around 800–1200 dots/inch, considerably better than the laser printers now available, but the methods used to drive them are similar. Figure 3 was produced on a computer-controlled laser typesetter with a resolution of 1000 dots/inch.

The effect of all these changes is to make it increasingly hard to distinguish between the hardware of a typesetter and that of an ordinary computer equipped with a high-resolution printer. In many cases they contain the same pieces of equipment hidden under different labels.

Dissatisfaction with current systems, together with this merging of hardware, is encouraging the creation of document preparation systems which provide the best features of the two worlds. Instead of discussing the latest systems in general terms, this paper tells the story of one particular system, giving a non-technical introduction to its facilities and explaining how it works.

Do not worry if you cannot understand every detail in the sections that follow. They are only intended to convey a feeling for the methods used and an idea of the power of the system. They are not meant as a complete guide to all the details.

The TEX system for technical text

In 1977 Professor Donald Knuth was told that there was no economic way of printing the second edition of Volume 2 of his book *The Art of Computer Programming* in the same style as the first edition. The book in question was highly technical, full of mathematical formulae and computer programs. Mathematical typesetting has always been unpopular because of its many special rules and the sheer difficulty of producing the necessary characters and special effects. It was said to have become uneconomic to have such a book typeset by the traditional methods and the computer-controlled systems could not, at that time, do the job properly. Professor Knuth is a perfectionist. He was not prepared to accept a lower standard, and promptly set about creating his own system to do the job. The result is the TEX system for technical text[3] and a beautifully printed second edition of his book.[4] The 700 pages of this book – mathematical formulae, computer programs, tables, running titles, footnotes, and all – were produced by TEX. Figure 7 shows a typical extract and some smaller examples are used in the following sections.†

The name TEX is taken from the same Greek root as English words like technology. It should therefore be pronounced as in TECHnology (*not* to rhyme with SEX). The name is intended to remind users that TEX is concerned primarily with technical manuscripts of the finest quality.

Preparing a TEX manuscript

To produce a document using TEX you need to create a 'TEX manuscript'. This is a computer file containing the text of the document interspersed with commands to tell TEX which character font to use and how to lay out the text. Every TEX command begins with the character '\' to distinguish it from the text of the document. (It is assumed that '\' does not occur in the text.) The word following the '\' must be one of the built-in 'command names' which have a special meaning to TEX. Sometimes further words or numbers are needed as part of the

†Knuth/Seminumerical Algorithms, The Art of Computer Programming, Vol. II ©1981 Addison-Wesley, Reading, MA. pp. 130, 216, 326, and 347. Reprinted with permission.

The failure of Cauchy's fundamental inequality

$$(x_1^2 + \cdots + x_n^2)(y_1^2 + \cdots + y_n^2) \geq (x_1 y_1 + \cdots + x_n y_n)^2$$

is another important example of the breakdown of traditional algebra in the presence of floating point arithmetic. Exercise 7 shows that Cauchy's inequality can fail even in the simple case $n = 2$, $x_1 = x_2 = 1$. Novice programmers who calculate the standard deviation of some observations by using the textbook formula

$$\sigma = \sqrt{\left(n \sum_{1 \leq k \leq n} x_k^2 - \left(\sum_{1 \leq k \leq n} x_k\right)^2\right) \Big/ n(n-1)} \tag{14}$$

often find themselves taking the square root of a negative number! A much better way to calculate means and standard deviations with floating point arithmetic is to use the recurrence formulas

$$M_1 = x_1, \qquad M_k = M_{k-1} \oplus (x_k \ominus M_{k-1}) \oslash k, \tag{15}$$

$$S_1 = 0, \qquad S_k = S_{k-1} \oplus (x_k \ominus M_{k-1}) \otimes (x_k \ominus M_k), \tag{16}$$

Figure 7. Extract from Volume 2 of *The Art of Computer Programming* (2nd edition, 1981) by Professor D. E. Knuth.

command. This sounds complicated but the examples given later in this section should make it clear.

The manuscript normally begins with a group of commands telling TEX about the overall appearance of the document – how big a page is, which character fonts are needed, information about page headings, how much the first line of each paragraph should be indented. There are a great many tedious details of this sort so, to save your time and temper, TEX comes with a set of default rules. You only need to give commands when you want something other than the default. If you require a specially small page size, 3 inches across and 5 inches deep for example, but otherwise you are prepared to accept all the default rules, then the start of your TEX manuscript would be simply

```
\hsize 3 in
\vsize 5 in
```

'hsize' and 'vsize' are two of the built-in command names, and, as the names imply, a size is also needed as part of the command. In this case the commands just tell TEX that the 'horizontal size' of a page is 3 inches, and the 'vertical size' is 5 inches.

You are now ready to give the text of your document. It does not matter how many words you type on each line of your manuscript. Unless you specify otherwise, TEX will extract words from the manuscript and fit them into output lines of the correct horizontal size with the left and right margins neatly lined up. TEX hyphenates words to avoid unsightly spacing, but tries to avoid this except where really necessary.

To prepare a straightforward document there are only six further commands you need to know about.

\par	indicates the end of a paragraph.
\noindent	cancels the normal indentation at the start of a paragraph.
\ctrline{. . . .}	causes the text between the braces to be positioned in the centre of a line.
\vskip	requests 'vertical' space between paragraphs. Space may be requested in inches (as shown in the \vsize example above) or in the traditional printers' units called points. (72 points = 1 inch.)
\vfill	requests enough vertical space to fill up the rest of the current page. \vfill is normally only used at the end of a document or at the end of a chapter.
\end	signals the end of the manuscript.

The example below shows the form of a complete TEX manuscript for a short document consisting of a single centred heading, followed by two paragraphs. The first paragraph is not indented.

```
\hsize 3 in
\vskip 18 pt
\ctrline{Fractions}
\vskip 18 pt
\noindent A picture in 'Punch' during the
1914-18 war showed an official saying to
a farmer, 'My dear sir, you
cannot kill a whole sheep at once!' \par
This absurd remark illustrates
```

```
the fact that fractions have no meaning
for certain things: you cannot
have half a live sheep. On
the other hand it is easy to have
half a foot of lead piping.
\par\vfill\end
```

The corresponding output is

Fractions

A picture in 'Punch' during the 1914–18 war showed an official saying to a farmer, 'My dear sir, you cannot kill a whole sheep at once!'

This absurd remark illustrates the fact that fractions have no meaning for certain things: you cannot have half a live sheep. On the other hand it is easy to have half a foot of lead piping.

Unless you specify otherwise, TEX will assume you want ordinary Roman characters in 10 point size. If you want another font, TEX has further commands to let you state your wishes. The commands \rm, \sl, and \bf cause a change to Roman, slanted (i.e. italic), and boldface characters, respectively. The commands \tenpoint and \eightpoint change the type size. Thus if part of the manuscript is as follows

```
Text is normally in Roman characters
but it can be \sl emphasized in italics \rm
or \bf highlighted in boldface. \rm The
size can also \eightpoint be altered like this
\tenpoint if desired.
```

The corresponding output is

Text is normally in Roman characters but it can be *emphasized in italics* or **highlighted in boldface.** The size can also be altered like this if desired.

TEX has a 'grouping facility' which avoids the need for quite so many commands. Text occurring between the braces '{' and '}' is treated as a single group. Commands occurring inside a group have no effect outside. Thus we could rewrite the above example:

Text is normally in Roman characters
but it can be {\sl emphasized in italics}
or {\bf highlighted in boldface.} The size
can also {\eightpoint be altered like this}
if desired.

Figure 8 shows the TEX manuscript used to produce the output shown in Figure 5. The example is chosen to show a variety of commands; a normal TEX manuscript would contain fewer commands. There are more elegant ways to achieve the same effect, but this example was produced by a complete novice (me!) on her first attempt at using TEX.

```
\vskip 18 pt
\ctrline{\bf I}
\vskip 18 pt
\ctrline{\bf Elementary Rules of Usage}
\vskip 18 pt
{\sl\noindent 1.  Form the possessive singular of nouns by adding ´s.}
\par\vskip 12 pt
Follow this   rule whatever the final consonant. Thus write,
\par\vskip 9 pt
{\sevenpoint Charles´s friend\par
Burns´s poems\par
the witch´s malice}\par\vskip 9 pt
Exceptions are the possessives of ancient proper names in
{\sl -es} and {\sl -is}, the possessive {\sl Jesus´}, and
such forms   as {\sl for
conscience´ sake}, {\sl for righteousness´ sake}. But such forms
as {\sl Moses´ Laws}, {\sl Isis´ temple} are commonly replaced by
```

Figure 8. The TEX manuscript used to produce the output shown in Figure 5.

How TEX works

Having given the reader a flavour of how TEX is used, it is time to look at how TEX sets about the job of setting type and forming lines and pages. As everyone knows, characters come in many different shapes and sizes. A lower case 'i' is naturally narrower than an upper case 'M'. Within a single character font there are many variations. When different type sizes and designs are taken into account, the possibilities are almost endless. TEX ensures that every character occupies its natural space; it also adjusts the spacing between words to make them fit neatly into lines with the same space between each word.

There are two further niceties of spacing that TEX takes

care of. One of these is the use of *ligatures*. Certain characters look bad next to each other and are therefore replaced by composite characters known as ligatures. The most commonly used ligature is 'fi'. If two separate characters are used the dot on the 'i' fights with the top of the 'f', so the two characters are merged into one. Figure 4 includes examples of ligatures.

The second nicety is the *kerning* together of certain pairs of characters. This means moving characters closer together than normal because of their shapes. Common examples are 'A' next to 'W' or 'V', 'o' next to 'x', and 'T' followed by a short letter like 'e' or 'y'. The rules for kerning and ligatures depend on the character font being used as well as the pairs of letters. A font with serifs (i.e. the little twiddly bits) will benefit more from kerning than a plain sans serif font – think of 't' followed by 'y'. The examples below show the effect of kerning (no kerning on the left, more kerning towards the right).

GO AWAY GO AWAY GO AWAY
Toxic Toxic Toxic

The TEX user decides which fonts to use and specifies the overall layout of the document, but TEX itself takes care of all the fine details of spacing, including ligatures and kerning. To do this it needs access to a library of detailed information about the fonts it is using. The following diagram gives a general picture of how TEX works.

TEX reads in the prepared TEX manuscript for the document and uses it, together with information from the font library, to decide exactly how the resulting document will look. What TEX outputs is a *device-independent* form of the document known as a DVI file. All spacing and page make-up is done by TEX, so the DVI file consists essentially of instructions to print given characters at given positions on the page.

To print the document, the DVI file needs to be converted into the necessary format for the printing device. This may be a laser printer, a typesetter or even a graphics screen. The program which does this conversion is known as the *device driver.* If the forms of the characters are built into the printing device, as they probably are for a typesetter, the device driver is a comparatively simple program. But if the printing device is a laser

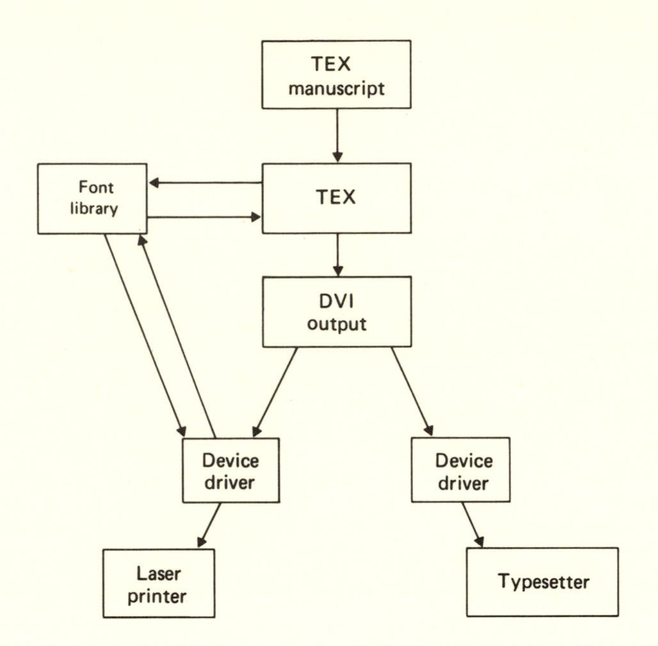

printer the device driver will need to consult the font library again to find out exactly what pattern of dots it needs to print.

Boxes, glue and pages

When laying out a line or a page, TEX is not concerned with the exact shape of each character, just its overall dimensions. It regards each character as a rectangular box with a base line running through it, as shown here for the letter 'q'.

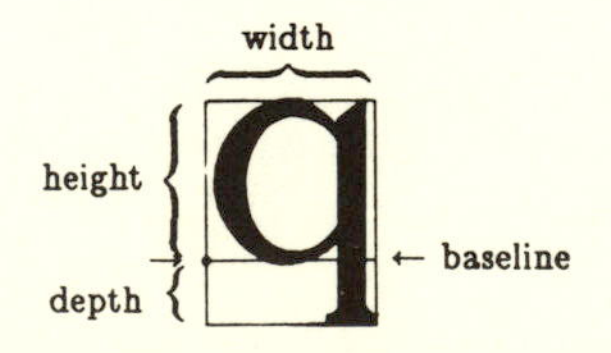

A word is made up by placing the boxes for each character in the word next to each other so that their baselines match up. Normally the boxes just touch, but they may overlap a little if any kerning is needed.

Words are then *glued* together to make up a line. Boxes are rigid. Glue is stretchable. Each blob of glue has a natural starting size and it also has the ability to stretch or shrink within given limits. TEX uses this flexibility to fit words onto a line. If a line needs to be stretched to fit its margins then each blob of glue gives a little according to how stretchy it is. It is traditional to leave more space after a punctuation mark than between two ordinary words, so TEX uses glue with more stretch but less shrink after punctuation. For each unit that ordinary interword glue will stretch, glue after a comma will stretch 1.25 units and glue after a full stop will stretch 3 units. The diagram below illustrates part of a horizontal list of boxes which is being glued together to form a line.

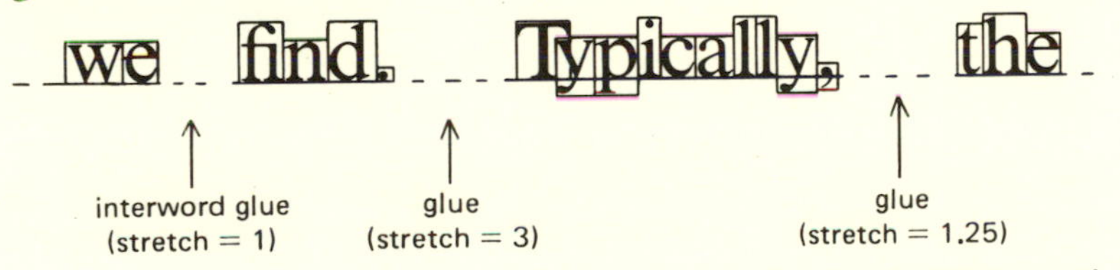

(Notice how the 'T' and 'y' have been kerned together.) The glue mechanism ensures that words are equally spaced along a line, except after a punctuation mark where the space is usually a little larger. Readers may find it interesting to go back at this point and compare the spacing in Figures 1 and 2 with the spacing produced by TEX in Figures 5 and 7.

When the glue has been set to form the final version of a line, the whole line is considered as a composite box (covering the tallest and deepest characters in the line). A page is made up in a similar manner to a line by glueing a vertical list of these composite boxes together and stretching or compressing the glue to make it fit the page depth.

Glue is very adaptable stuff. A fixed size space can be provided by glue with zero stretch and shrink. At the other extreme, the \ctrline command uses glue with infinite stretchability at either end of the line. Any spare space is thus divided equally between the two infinitely stretchable blobs leaving the text untouched in the centre of the line.

TEX takes a great deal of trouble to find the best possible layout for a document. Two fundamental decisions are when to start a new line and when to start a new page. Many systems will only consider a line at a time, but TEX always considers the layout of a paragraph as a whole. It will also look a long way ahead before deciding on the best place to start a new page. A system of penalty points is used to decide how good or bad a particular layout is. The best layout is the one with fewest penalty points. Examples of problems leading to penalty points are:

(i) overstretched or overcompressed glue;
(ii) hyphenation at the end of a line (hyphenation at the end of two or more consecutive lines leads to an enormous number of penalty points);
(iii) 'widows' or 'orphans' (i.e. leaving the first or last line of a paragraph isolated because of a page break);
(iv) line or page breaks in mathematical formulae.

In addition, the user may specify penalties for line or page breaks by inserting a \penalty command at the critical point. A penalty of 1000 or more points means that a break is forbidden at that point under any circumstances. A negative penalty, on the other hand, indicates a desirable break point.

If you find this description of TEX's way of making up pages difficult to follow, you may prefer Professor Knuth's way of putting it. He simply says: 'Conceptually, it's a big paste-up job'.

Advanced facilities of TEX

It is impossible to cover more than a tiny fraction of the facilities of TEX in a short paper such as this – the TEX users' manual runs to 200 pages – but there are two major areas which simply cannot be left out altogether.

The first of these is mathematical typesetting. TEX knows all the traditional rules for printing mathematics, and has a special 'maths mode' where it deals with unusual symbols and spacing requirements. It will, for example, automatically change to smaller type for subscripts (and to still smaller type for subscripts of subscripts and to still smaller type . . .). Anything appearing in a TEX manuscript between pairs of dollar signs is taken to be in maths mode. The following examples show how mathematical formulae can be built up. Each example shows what you need to type in the TEX manuscript to produce the corresponding output.

(i) `$$t↓2+t↓1+3t↓3=4$$` $t_2 + t_1 + 3t_3 = 4$

When you use a ↓ character in maths mode it means that the following item (or group) is to be printed as a subscript. Note the different sizes and positions of the two instances of the number 3 above. TEX uses standard mathematical spacing and therefore ignores any spaces typed in maths mode. Exactly the same output would be produced by '`$$t ↓ 2 +t↓1+3t↓3 = 4$$`'.

(ii) `$$(k+x)↑2$$` $(k+x)^2$

This example is similar to (i) but uses ↑ for a superscript.

(iii) `$$ 1+x \over (k+x)↑2 $$`

$$\frac{1+x}{(k+x)^2}$$

In this example \over tells TEX to place one item over another. TEX works out the length of the horizontal line needed and centres the items above and below to look neat.

(iv) `$$h(x) = \sum↓{k≥1} {1+x\over(k+x)↑2}$$`

$$h(x) = \sum_{k\geq 1} \frac{1+x}{(k+x)^2}$$

This example shows a more complex formula being built up. Note the use of grouping with { and } to ensure that \over only applies to the items within the group.

(v) `$$X↓1↑2+X↓2↑2+ \cdots +X↓n↑2$$`

$$X_1^2 + X_2^2 + \cdots + X_n^2$$

\cdots is used to produce the dots.

(vi) `$$r=\sqrt{X↓1↑2+X↓2↑2+\cdots+X↓n↑2}$$`

$$r = \sqrt{X_1^2 + X_2^2 + \cdots + X_n^2}$$

This is another example showing how complex formulae can be built up from simpler pieces. \sqrt produces a square root symbol covering everything in the following item or group. If the { and } were omitted in this example, the square root symbol would only cover the first X.

These examples are necessarily sketchy, but readers should now understand how some of the formulae in Figure 7 were produced, and really intrepid readers may even be able to work out how to produce something as elaborate as

$$\lim_{n\to\infty} \frac{\frac{1}{n}\sum U_j U_{j+k} - (\frac{1}{n}\sum U_j)(\frac{1}{n}\sum U_{j+k})}{\sqrt{(\frac{1}{n}\sum U_j^2 - (\frac{1}{n}\sum U_j)^2)(\frac{1}{n}\sum U_{j+k}^2 - (\frac{1}{n}\sum U_{j+k})^2)}} = 0.$$

Macros are the last feature of TEX to be considered. Macros provide a means of defining new commands in terms of existing commands. At the simplest level they provide a convenient shorthand notation, but at a higher level they provide an important means of tailoring the system to a particular job. Like many complex features, macros are best explained by looking at a typical example.

Suppose you are preparing a TEX manuscript for a long document containing several Chapters subdivided into sections. Each section is introduced by a heading. Initially you decide that section headings should be printed in bold type and should be surrounded by a certain amount of white space to make them stand out from the rest of the text. This can easily be achieved by typing the relevant commands directly into the TEX manuscript for each heading, for example

```
\vskip 36 pt
{\bf.................}
\vskip 18 pt
```

where the dots represent the text of the heading. This is rather tedious to type for a large number of section headings, so you decide to define a new macro command called, say, \sect which reduces the typing for each heading to

```
\sect{...............}
```

To do this you include a definition of the new macro at the start of your manuscript. The definition for \sect is

```
\def\sect#1
{\vskip 36 pt
{\bf #1}
\vskip 18 pt}
```

This tells TEX to augment its set of built-in commands with a new macro command called \sect, and to define the new command to be equivalent to the sequence of three commands

```
\vskip 36 pt
{\bf #1}
\vskip 18 pt
```

where the item or group following \sect in the manuscript is to appear in place of '#1'. The definition of the macro need only be given once, but the macro can be used any number of times. It thus achieves its original aim of providing a convenient shorthand notation at the expense of a very small overhead.

As your document progresses two things will happen. First, you will find that unfortunate page breaks can occur, leaving your section headings at the very bottom of a page. This can easily be cured by using \penalty to encourage TEX to start a new page immediately before a section heading and to forbid it to start a new page immediately after. If you are using the \sect macro to start each section, you only need to change the definition of \sect to

```
\def\sect#1
{\penalty -200
\vskip 36 pt
{\bf #1}
\penalty 1000
\vskip 18 pt}
```

Second, you will become more ambitious and will want to experiment with different layouts and complex page headings. If you have made a sensible use of macros in your manuscript you can work wonders by changing a few macro definitions. The definition of \sect could easily be changed, for example, to use larger type for the heading. With a little more skill it could also be made to insert the section heading into the running title of all subsequent odd-numbered pages.

The **\sect** example was chosen to show that macros can provide much more than a shorthand notation. They help to ensure consistency because every call of a given macro is guaranteed to be equivalent to the same set of commands. Inconsistencies due to omissions or typing errors are thus reduced. Macros also make it easy to change the format of a long document. But, perhaps most important of all, they allow the user to prepare a document manuscript in his own terms. Documents are usually made up of a small number of constituent parts. Typical examples are chapters, sections, subsections, numbered lists, footnotes, and perhaps less common items like theorems, tables, and references. If the user defines macros to cope with the layout of the basic constituents of his document – and disciplines himself to use these macros whereever they apply – then his TEX manuscript immediately becomes more understandable and all the benefits of consistency, ease of change, and conciseness follow.

Current status of TEX

TEX's original goal has already been achieved with the publication of Professor Knuth's book, but its story certainly does not end there. So many people became interested in the power and quality of TEX that a 'portable' version has been produced. This is already working on eight different types of computer and is being implemented on several more. Similarly, TEX output can be sent to an increasing range of printers and typesetters. The thriving TEX Users' Group publishes a newsletter[5] containing progress reports, details of TEX implementations, and ever more complicated packages of macros to make TEX do ever more complicated things.

Most of this activity comes from the research and academic world. But there are signs that TEX may already be bridging the gap to the commercial printing world. One sign is that the American Mathematical Society is experimenting with TEX for publishing its journals. Instead of submitting the normal printed version of a paper, the author can now submit a magnetic tape containing a TEX manuscript. The American Mathematical Society expects this to save 50 to 80 per cent of normal publishing costs.[6]

The second sign is that manufacturers of commercial typesetters are becoming interested. The main difficulty in providing a device driver to print TEX output is the compatibility of character fonts, especially mathematical fonts. For his book Professor Knuth developed a family of fonts called *Computer Modern.*[7] These can be guaranteed to form part of the font library for every implementation of TEX and have come to be regarded as the standard TEX fonts. It seems likely that they will shortly be made available by several manufacturers so that TEX output may easily be printed on their typesetters.

Other systems – SCRIBE and BRAVO

TEX is outstanding for mathematical typesetting and for high quality printing. Different systems, however, have different aims[8] and it is instructive to look briefly at two systems with no mathematical abilities.

The SCRIBE system[9,10] is similar to TEX in many respects. Both are 'batch' systems which require a complete manuscript for a document to be prepared in advance. The user then feeds the manuscript to the system, but has no further interaction with it while the manuscript is being processed. Like TEX, SCRIBE copes with different character fonts and produces output which can be sent to a variety of printers. The most distinctive feature of SCRIBE is its 'cookbook' approach – it contains built-in 'recipes' for a variety of document types. A SCRIBE user does not normally need to bother about fine details of layout. Instead, he looks for a suitable recipe containing all the ingredients needed for his document and uses the very simple built-in commands to output those ingredients. One of the simplest document types is a business letter. Typical ingredients of a letter are:

– an address
– a greeting
– the body of the letter
– a closing phrase

A more complicated document, like a thesis or a manual for example, may contain all the following ingredients

– a title page
– a preface
– a table of contents
– chapters
– numbered sections
– references
– appendices

The first two commands in a SCRIBE manuscript give the type of document and the name of the printing device to be used. The ingredients for that type of document then follow in order. An outline of a SCRIBE manuscript for a thesis to be printed on a Xerox Dover printer is

```
@Make(thesis)
@Device(Dover)
@Begin(TitlePage)
........ text of title page .....
@End(TitlePage)
@Chapter(... title of first chapter ...)
...................
@Section(... section title ...)
...................
@Section(... section title ...)
...................
@Chapter(... title of next chapter ...)
...................
...................
@Appendix
...................
```

SCRIBE automatically numbers chapters and sections and it has excellent facilities for producing a table of contents and an index.

If TEX is supplied with suitable packages of macros it can be made to appear very much like SCRIBE. The difference is one of basic philosophy. SCRIBE is designed to be simple for non-technical users who want good quality output but who are prepared to accept a limited number of standard layouts. It is possible to force SCRIBE into different ways but that is to run contrary to its nature.

The BRAVO system[11] adopts a completely different approach

to creating a document. It relies on the use of a high-resolution graphics screen and a pointing device called a 'mouse' which allows the user to move a cursor around on the screen. Creating a document is an interactive process where the user sees the document taking shape on the screen as he types it in. The current state of the document is what he sees on the screen; there is no separate 'BRAVO manuscript'. When the user is satisfied that the document is correct he issues a command to print it. A system of this sort is sometimes called a 'you-get-what-you-see' system.

To create or change a BRAVO document the user first selects the position to be changed and then types a command specifying the change. A position is selected using the mouse which is a small metal object running on a ball bearing; it has three buttons on top. Pushing the mouse around on a flat surface spins the ball and, through a marvel of electronics, causes a cursor to move around on the screen in unison with the movements of the mouse. When the cursor reaches the desired position a button on the mouse can be pressed to select that position; BRAVO then displays a small blinking mark under the character selected. To insert some text at the selected point, the user simply types I (for Insert) followed by the text to be inserted. As each character is inserted any following text is moved up to make room for it – words will overflow onto the next line as necessary. The user sees this happening character-by-character as he types the new text in.

A whole chunk of text can be selected (by selecting its first and last characters) and commands given to alter the entire chunk by changing it to a different character font, for example, or by indenting its left-hand margin. Again, the user sees the effect immediately. One particularly useful command is U (for Undo) which simply cancels the last change made. If the user makes a change and, on seeing its effect, decides that the original was better after all, he can Undo the change immediately.

This interactive method of working is delightful. Errors can be spotted immediately and it is easy to experiment with different layouts. For short documents it is undoubtedly the ideal approach, but it has one serious drawback for long documents. Because there is no separate BRAVO manuscript there is no way of making systematic changes to the document by changing a macro definition. If the user decides to change all his section headings he needs to search through the entire document and change each one individually. A lot of research needs to be done to find the best way to merge the delights of interactive document creation with the power of a batch system.

Distributed computing and typesetting servers

Changing economics and advances in computer communications mean that computing is moving away from the age of the massive mainframe computer into the age of the distributed system where smaller computers are linked together into a local area network. In the ideal distributed system each user on the network has his own 'workstation' consisting of a moderately powerful personal computer, complete with high-resolution screen and some local disc storage. The personal computer is powerful enough to handle most routine tasks and gives the user far quicker service than he would get if he were fighting for a share of a large mainframe computer. When he needs something beyond the powers of his workstation the user can call on a range of services provided elsewhere on the network. With this kind of distributed system it is not necessary to provide specialised facilities on every computer; instead a single 'server'

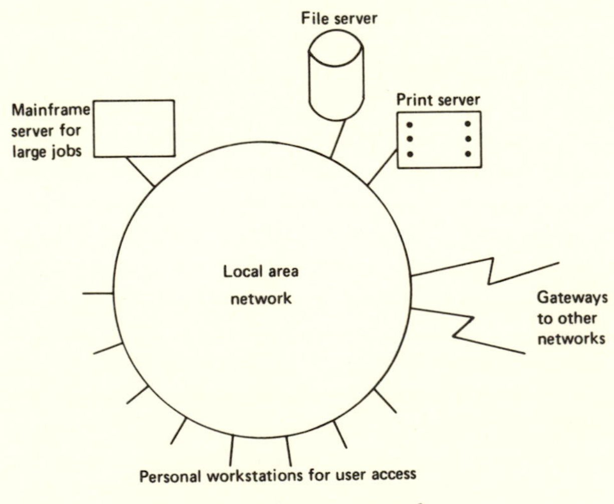

Figure 9 Local area network

can be provided and made accessible to every user on the network. Figure 9 shows a typical local area network consisting of workstations, servers and 'gateways' to other networks. Servers may include mainframe computers to provide the sheer size and power needed for large jobs, but a typical server is a smaller computer managing the use of equipment such as discs, plotters or printers. A file server, for example, is equipped with large discs and manages them as a large and secure filestore to augment the limited facilities at each workstation.

This distributed approach is ideal for a document preparation and typesetting service. To provide a good service at least one high quality printer or typesetter is needed and a lot of specialised software. It would be wasteful and prohibitively expensive to provide these on every computer. Instead, everything is concentrated into a single dedicated typesetting server, thus providing a high quality service to a large community of users at a moderate cost.

Conclusion

Shoddy printing is not going to disappear overnight, but there is no doubt that better printing devices and better document preparation systems exist and are spreading rapidly. Once computer users have had access to a high quality printing service they will never be satisfied with lineprinter output again. A great many office systems are already distributed systems of one sort or another[12] so the networks exist too, and it is only necessary to plug the typesetting servers into the networks to reverse the flood of badly printed documents.

The whole printing and publishing world is in a state of flux as new computer systems replace old equipment. Many experiments are under way to match the document preparation system of the programmer to that of the printer. The resulting marriage should be the best thing since a similar marriage of the baker and the cutler produced sliced bread. Indeed the outcome should be far superior to the insipid product of the baker and the cutler.

References

1. Teitelman, W., 'A display oriented programmer's assistant', *Proceedings of the 5th International Joint Conference Artificial Intelligence*, Vol. II, pp. 905–15, 1977.
2. Phillips, A. H., *Handbook of Computer-Aided Composition*, New York, Marcel Dekker, 1980.
3. Knuth, D. E., *TEX and METAFONT: New Directions in Typesetting*, Bedford, Mass., American Mathematical Society and Digital Press, 1979.
4. Knuth, D. E., *Seminumerical Algorithms – The Art of Computer Programming*, Vol. 2 (2nd edition), Addison–Wesley, 1981.
5. *TUGBOAT – the TEX Users Group Newsletter*, published irregularly by the American Mathematical Society, Providence, R. I., 1980/82.
6. Swanson, E. E., 'Publishing & TEX', *TUGBOAT* 1 (1), 1980.
7. Knuth, D. E., *The Computer Modern Family of Typefaces*, Stanford University Report No. STAN-CS-80-780, 1980.
8. *Proceedings of the ACM Symposium on Text Manipulation*, published as *SIGPLAN Notices* 16 (6), 1981.
9. Reid, B. K., *A High-level Approach to Document Formatting*, Record of the 7th ACM Symposium on Principles of Programming Languages, 1980.
10. Reid, B. K. and J. H. Walker, *SCRIBE User Manual*, UNILOGIC Ltd, Pittsburgh, PA., 1980.
11. Lampson, B. W., *BRAVO Manual*, Xerox Palo Alto Research Center, 1979.
12. Naffah, N. (ed.), *Integrated Office Systems – Burotics*, North-Holland, 1980.

8. HUMAN COMMUNICATION AND INFORMATION TECHNOLOGY

PHILIP J. HILLS
Primary Communications Research Centre,
University of Leicester

As the new technologies become more and more the technology of the present and not the technology of the future, more concern is being shown for what has been called the 'man/machine interface'. In the context of this discussion the basic components of the new technologies, i.e. the 'machine' side of the interface would seem to be usefully summarised as the computer, often with network linked databases, and the video screen or high speed printer for output and display of material.

Much concern has been shown for the ergonomic factors of comfort – ease of use of keyboards, comfortable seating, prevention of eyestrain in the prolonged use of screens, etc. Some concern is being shown for the use of these systems not just in terms of their presentation of information but in terms of the need for this information to be in a form that is easily useable and easily assimilated. I believe that one needs to look at the needs and purposes of the user of information in some detail and then to design systems which actually help the user rather than simply present information and ask him to adjust to it.

If we consider the man/machine interface then the brain of man can be considered to be a vastly powerful biocomputer and the machine is an electronic computer. Each of these computers offers specific advantages, the electronic computer has tremendous potential for fast and accurate recall of data, the biocomputer possesses immense flexibility and adaptability.

From this it would seem that the present position where the electronic computer displays data which the biocomputer adapts to is a good use of the advantages of both systems. However, this neglects the important point that although the electronic computer can be built to be more and more powerful, as yet, human biocomputers come in a variety of types and sizes.

What is not yet clear (the nature/nurture debate) is whether it is merely faulty programming on one hand and more effective programming on the other that creates the differences in flexibility and adaptability in different human biocomputers. This is not an argument that I wish to pursue here but merely to note that these differences do occur and hence recognise the need to take these into account when examining strategies and possibilities.

In this situation therefore we shall consider the information system representing the new technology as the constant factor, capable of delivering data on a random access basis, at high speed from a supposedly unlimited source. The variation is thus in the processing ability of the human biocomputer. To avoid unnecessary usage of the term human biocomputer I shall now revert to the term 'user'. We must also recognise that in the context of this paper we are talking about a user who is a professional of some kind who has most likely acquired skills of accessing and retrieving information using textbooks and conventional library sources. In the course of his training he may well have received some instruction on 'study skills'.

Typically, study skills advice consists of advice on how to organise your time, how to take notes in lectures and so on. However, part of that advice is concerned with how to use books, how to take notes from books and how to use information to prepare essays, reports, etc.

Study skills advice thus aims to prepare students to be flexible and adapt to conventional sources of information, since books by their nature are essentially inflexible information providing devices. Textbooks often consist of predigested information in some form or another and are sufficient for many purposes, as to go back to the primary source of the ideas contained in them would often be difficult and costly in time and money. Information systems at present in use often follow the textbook or encyclopedia in concept and consequently have adopted the same premise, namely that the user will adapt to the presentation instead of the other way round.

The programmed learning movement which was at its peak

in this country in the 1960s was an attempt to break the inflexible linear presentation of text. It was an attempt to help the user to assimilate and understand the material presented by using a programmed question and answer style rather than the normal book format. Although the programming methods that programmed books attempted to use were difficult to design for print on paper methods they are easy for devices like the microcomputer. They did, however, represent some attempt to process text so that it attempted to be flexible and adaptable to the user. My point is that computers are very capable of such flexibility and we need to think of them more in this way.

In order to illustrate some ways in which this flexibility might be possible let us take the example of someone who is required to prepare a paper in a specialist area. This consists of the process of aquiring, using, transforming and sending out information.

The process starts with the idea for the paper and the intending author decides on the approach in terms of its subject matter, scope, intended audience, and level of treatment. The stages outlined below are obviously not the only way of going about the preparation of a paper, and a set of tasks like this may not even be explicitly stated by the author, but even so, to some extent they are there.

Acquiring information

The author draws upon background knowledge to answer a series of questions such as the following:

1. What do I know that is relevant?
2. What else do I need to know to cover the subject adequately?
3. Where do I know that I can go for other relevant material?
4. Where can I find out where else to go?

Using a version of the technique that I have often advocated, known material can be written on cards so that it can later be restructured. The answer to 'What else do I need to know?' can also be written on cards as a series of keywords to be filled out from other information sources later. When all information is available the cards can then be shuffled into a logical order and the cards used in turn to aid writing the paper.

The main problem to be brought out here is the need to search information sources using keywords or phrases to find papers and articles which represent a coverage of the subject area. At present regardless of whether this search is done manually using a printed abstracting service or a computer database it would be necessary to check the range of keywords used by the abstracting service and to choose those which conformed as nearly as possible to those that represent the ones chosen for your search. The emphasis here, as usual, is on the user to conform to the requirements of the service offered.

It would not be impossible to build up a thesaurus of near equivalent terms for each database so that the user could input his own keywords and these would be translated into near equivalents which would be used to search the database. This would be a relatively simple way to place the need for flexibility and adaptability on to the electronic computer.

Using and transforming

Given a successful search using relevant keywords, details of a number of articles will now be generated. Conventionally the only way to judge their relevance is either to access them in one's own or another library, or to obtain them on the inter-library loan system. The electronic alternative would seem much quicker and more attractive. If it were possible to call up the texts on a screen these could be scanned, absorbed and used for the author's present purpose. There are at present however a number of problems in accessing this material from a screen. Conventionally a lot of paper and hence a lot of reading can be generated by this process much of it unnecessary and unneeded. When this is translated on to a screen it becomes even more difficult and tedious as long reading sessions using present video screens is not a thing to be contemplated. We are promised better graphics, better text and high resolution screens, less glare, etc. and this will certainly improve the readability of such material. It does not, however, get rid of the problem of sifting the material needed from the material presented. To aid this problem authors are going to have to learn more discipline in their use of an agreed system of keyword subheadings that will enable specific parts of an article to be accessed easily and accurately. Since authors are readers and many readers may be authors then the need for this should be very apparent when full text becomes available by recall from databases. Maurice Line's article in this volume outlines with examples ways in which articles could be designed for on-line viewing.

The process of transforming the raw material into the final paper can be aided by information displayed to the author on cards, paper or screen, but the process goes on largely within his brain. Here the new data presented is used to extend existing data, it is compared, measured and recorded. This results in a series of statements which, for example, agree with or disagree with, extend or modify previous work or offer alternative explanations or solutions. This would not be an impossible task for an electronic computer but one that I would venture to suggest would be difficult and indeed unnecessary for it to perform.

Output

The material is then transformed into output. Material is ordered into a logical sequence, written out in draft form, polished, edited and then made ready for publication. The final process of ordering the material can take place using cards which can easily be moved about on a table until the desired sequence is obtained. This method also has the advantage that it exposes gaps in coverage for which other material can then be obtained.

This process could readily be performed on a computer. If each section of material was typed in, using a keyword or phrase as identification, then a display of the keywords could be called up after all the material had been put in. These keywords could be moved about in the list until the sequence was acceptable. At this point the original material could be called up in the sequence decided upon. This of course would give no more than the summary information and it would also be necessary to display the final version of the paper as it was keyboarded in. This could be done by holding and displaying the summary information on one half of a split screen with the final version on the other. The limitation here is of course in the amount of information displayed. This could be overcome by the use of a large projection screen where several pieces of information could be displayed together.

Regarding the need to accurately access parts of articles when full text becomes available on-line, even papers which present primary data may contain an amount of material which merely duplicates other material. Take, for example, a paper describing advances in a particular area. This will probably outline events and important pieces of work leading up to work described in the paper. Not only can this lead to unnecessary duplication if all material published in this area is easily and quickly available, but also a review in one article may well be incomplete and selective. This will mean that if the transmission of primary information by computer and linked databases becomes the dominant method, then authors will have to develop new skills of writing, or perhaps no skills at all except those of being able to present the design and results of their experiments in as plain a form as possible. However, since they will all have access to all previous primary data on the subject via the database, all that is needed is knowledge of how to access key papers which preceeded their work, together with a justification and presentation of the problem as they saw it leading to the experiment which is now being presented.

The processes of preparing the final draft, editing the paper and preparing it in a form that can be used for the final method of transmission, be it print on paper or for storage in a computer database, are available now and need no further elaboration here.

The example of the stages in the preparation of a paper has been used to illustrate the possibilities in making the electronic computer bend towards the user. There are many other examples and possibilities of this and some are described in Paul Lefrere's article which follows this one. The answer, inevitably, lies somewhere in the middle, somewhere between the present inflexibility of the presentation of information by computer and the varying degrees of flexibility than are capable from the user. One thing however is certain, and that is that over the next decade the use of electronic information systems will increase. This paper is a plea to the designers of such systems to take account of the needs of the users so that they can be actually helped rather than hindered in their information gathering tasks.

9. BEYOND WORD-PROCESSING: HUMAN AND ARTIFICIAL INTELLIGENCE IN DOCUMENT PREPARATION AND USE

PAUL LEFRERE
Textual Communication Research Group,
Institute of Educational Technology,
Open University, Milton Keynes, UK

Introduction

The title of this chapter needs some explanation, which I precede with two observations. First, like the Laputan writing machine in Swift's *Gulliver's Travels*, conventional word-processors make it easy to produce documents which are not used or which do not achieve their purpose; a common example is the 'personalised' mail-shot. Second, like Winston Smith (the human word-processor in Orwell's *1984*), a word-processor can search for specified phrases and replace them, yet still requires supervision to ensure that this operation is going to plan. The reason in the machine's case is its lack of *intelligence.* My chapter title is meant to reflect both how this is changing and how our ideas of word-processor design and use are changing.

Cryptologists were among the first to realise that computers can remember and manipulate words as well as numbers. By the 1960s this idea had percolated to the office, together with terms such as 'information technology' (Heyel, 1969). What has changed since then? Certainly, computers have become a more common sight. They are cheaper, faster, easier to use and easier to connect to other computers and machines. But accompanying these changes in computer technology have been changes in our understanding of human skills and capabilities.

Although we are still a long way from the 1950 vision of 'Machines that Think and Want' (cited in McCullogh, 1965), it is technically possible for computers linked to word-processors to model or simulate many aspects of document preparation (including doing the writing themselves!). For example, they can act as intelligent clerks, tutors, advice-givers, assessors or editors of our work. This is partly what this chapter is about. It is also about what we humans do when we settle down to write, to type, to edit or even to read – all important aspects of information transfer! While the issues are technical, I have tried to pitch the discussion at a level suitable for beginners as well as experienced information technologists. As an earnest of this, the next section provides a brief primer on word-processing.

The what and why of word-processing

Both data-processing and word-processing are based upon digital computers. Both involve the storage and manipulation of information. The difference is that in the simplest form of word-processing the information is in the form of language rather than numbers. To be more precise, it is in the form of alphanumeric characters – the alphabet, numerals and other symbols we use when we write. More advanced systems accept written and spoken annotations and some extend the idea of word-processing to illustrations, for example letting us 'draw' diagrams using the technique of selecting and positioning graphic 'icons' from a kind of palette – a computer menu. They even permit us to design our own typefaces, mixing several, of different size, within the same document. More common are slightly less powerful machines which can be used for calculations or for constructing an in-house database, as well as word-processing. Most of these can be linked to other machines to send messages or documents, or even to access large public databases. They can then accelerate or augment our ability to retrieve and process information or even have tasks delegated to them, depending on how the information is stored and how well we can specify the tasks (Lefrere and Whalley, 1979).

However complex the information stored, we can move it from one document to another or from one word-processor to another. There are many ways of doing this and several technologies for doing it. For example, in the fairly near future we may combine electronic word-processors with optical computers (a non-electronic technology) or microfilm (a non-computing technology). Such details of the technology and the systems available are continually changing so I deliberately devote little space to them.

A typical basic word-processor has a typewriter-like keyboard, plus some unfamiliar keys used to operate the machine. Its prime function is to record the sequence of key depressions as a document is typed. That sequence can then be recalled later for amendment or for re-use with no need to re-type. Thus, even the most basic word-processor has several common uses, uses which would otherwise involve repetitive typing. These range from work with a standardised, repeatable element, such as the assembly of letters from a stock of pre-typed paragraphs, to the re-drafting of an out-of-date or incorrect document by making insertions and deletions. Its ability to help with the mechanical side of re-drafting makes it attractive to authors, some of whom now send a word-processor 'floppy disc' record of their manuscripts to their publisher, together with the usual typed copy. In this lies its interest to publishers: like optical character recognition, it may help to contain the costs of unnecessary re-keyboarding.

From word-processing to the memex

Even the simplest word-processors can be used for information retrieval, although they will not be as effective as a purpose-built retrieval system. As word-processors and microcomputers become cheaper, it becomes feasible to think of using them as a component of a modern-day version of the 'Memex' – a pre-computing, personal-information-system (Bush, 1945).

For those unfamiliar with the Memex concept, Bush thought of it as a desk-sized machine with mechanically linked microfilm storage and copying facilities. Users would store in it copies of books and conventionally published papers, as well as their own notes and records. It would allow both rapid searches through the stored material (for retrieval, sorting and annotation) and the addition of new and unrelated material. Its purpose was to improve information flow, originally in science: while major contributions to a field would continue to be published in the conventional way (with copies stored subsequently in the Memex), minor contributions would be circulated to Memex users only.

Bush's idea is of interest both because it describes a model of document preparation which continues to influence much modern thinking and because it concentrates heavily, in my view, on dissemination. The idea was to introduce some form of 'gatekeeping' to reduce unnecessary reading. Simplifying Bush's argument considerably, the typical minor contribution might be for the most part a recompilation and annotation of documents already in the Memex, a merging of a diagram from document A and a paragraph from document B. Its 'publication' would require no more than giving details of the *trail* taken through the common stored material ('read this paragraph then look at that diagram') together with details of any *annotations*.

Like many grand ideas, Bush's speculations were so influential that subsequent workers concentrated on how to implement this idea, rather than asking whether it was really appropriate for authors and readers. For example, would it help authors to overcome the writing hazards of 'cold start' (where and how to begin writing) and 'writer's block' (being unable to express ideas as text)? Similarly, would readers find it as easy to browse through as a book? Current research indicates that readers might lose their way, forget where they had been and where they were going and perhaps find the 'joins' between one screen's worth of information and the next too distracting (see Robertson et al., 1981; Fox and Palay, 1982). Powerful methods of finding out what is needed have been developed in cognitive science and other disciplines. As we see later, this evaluation can proceed either at a fine level (e.g. interpreting sequences of key depressions or content-analysing the types of contributions made and the 'trails' taken) or at a gross level (e.g. looking at the 'take-up' by authors of opportunities to publish via a Memex network, or the effect of such a network on editors and others).

One very gross measure of appropriateness is whether the system is found to exert psychological stress on users (see Senders, 1980). To put this another way, we must not forget that computer systems do not terminate at their terminals, so we must consider the user before changing any part of a working system:

> . . . new inventions [such as the high-resolution, page-size display of expensive word-processors] introduce their own problems . . . [including] display-management problems

> that did not arise with the teletypewriter terminal: what information should be displayed? How should it be displayed? How can we be sure that the user sees a message? (Moran, 1981:2)

So to say 'we have the technology' is not enough; we require a broader perspective of what it is that users do or attend to, and why users' behaviour is as it is. The need for this was well expressed in the companion volume to this, *The Future of the Printed Word*:

> no matter how the information is transmitted and displayed, . . . people will *still be reading.* If we really wish to improve *communication*, then, we must attend to much more than the tools whereby words are recorded, moved from one place to another, and presented to a consumer of information. . . . We need to understand the determinants of comprehension and retention, and to put that understanding into practice in the *creation* of the communications. . . . In other words, we need to learn a great deal more about . . . the human engineering of communication' (Strawhorn, 1980: 24–5).

Human factors

Providing information or making decisions on office automation may be seen as the province of particular groups – for example, accountants or organisation and methods units. If those groups do not appreciate the details of the tasks being undertaken or do not draw upon what is known about work organisation, social groupings, etc., then this automation can lead to a reduction in efficiency. It can also lead to a lessening in the satisfaction people derive from their work. Yet acquiring pieces of new technology need not be equated with – as some say – 'replacing intelligent living labour with the dead'. It is a great pity (but only peripheral to this chapter) that the man-years invested in the equipment side of computer technology have not been matched by an equivalent amount of research devoted to making fuller use of human abilities and enabling more of us to contribute to society. It is also unfortunate (and *is* a matter for this chapter) that making machines such as word-processors easy to use has been the subject of less research than making them cheap to buy. In both these cases there *is* a body of research worth noting, but this is not yet well known because it is spread over a number of academic disciplines; the major ones are: ergonomics (known as 'human factors' in the USA), cognitive science and artificial intelligence. (Since the latter two may be unfamiliar, the next section gives a very brief description of each.)

Although not couched in this way, the available research indicates that to build a system which is as simple and unintimidating as the telephone, we must satisfy certain criteria, including:

- the avoidance of warning lights and error messages;
- the provision of help through the system itself (a machine equivalent of Operator Enquiries, perhaps); and
- the design of a system which can be understood quite well enough to use it without technical knowledge (analogous to thinking of the telephone as a speaking tube?)

If designers fail to do this, then familiar tasks such as writing or editing may have to be carried out in unfamiliar and possibly less meaningful ways. Training programmes can then still leave users making errors or prone to feelings of insecurity (e.g. worrying about what happens to the words you type as they go off the screen of your word-processor). Early typists, even of such complicated machines as the original IBM composer, may have been helped here because all parts of their machines were exposed. Later machines had fewer parts exposed, but these and their actions were still visible through glass windows set into the sides of the machines. Modern word-processors and computers don't have the equivalent of windows, and what goes on inside them has no obvious parallel with things we know about. Only with a few, very expensive machines do we find instructions presented in everyday terms (e.g. pictures of in-trays and out-trays, which we can 'point' to on the screen to inspect their contents).

To restate these problems in the jargon of cognitive science: before we can be happy in our use of a machine, we need a good 'mental model' of what it does and of the information which comes from it. This need not be strictly accurate; young children being taught to program may be better served by using concrete analogies such as 'men passing messages inside the computer' than by giving them actual (and more abstract) technical details of what the instructions tell the computer to do. Without analogies or some other means of placing instructions in an understandable context, they may seem unfamiliar and arbitrary:

> Place two fingers in the two holes in the disc directly to the left of the protruding bar. Remove finger nearest protruding bar. Rotate disc to protruding bar. Remove finger and allow disc to return. Repeat operation twice more.

Substitute for 'dial' 'disc' and 'finger stop' for 'protruding bar' and you may recognise these as the telephone directory's instructions on how to dial 999 in darkness or smoke. The sequence of instructions is short but it lacks a familiar pattern for most readers. Even where the context seems clear enough, we may bring to a machine some assumptions which its designer does not share. An everyday example would be a telephone system in which the engaged tone meant 'wait – no need to re-dial', rather than 'try again later'. Another, word-processing example is the meaning you attach to 'delete' and 'erase'. On some machines, both have the same meaning. On others, the designer's intuition has told him or her that you place a *different* interpretation on each (which you do, but not *his* interpretation!), or that you understand what he is doing when he changes the meaning of each term from one task or context to another. But designers can do better than this if they take account of recent research, particularly in cognitive science.

Cognitive science and artificial intelligence

The terms 'artificial intelligence' (AI) and 'cognitive science' have been used several times already without any explanation. One is therefore overdue. The terms refer to twin disciplines which go back about two decades. Although they are likely to have a considerable influence on our ideas of information transfer, their technical nature and relative newness mean that few members of the public know much about them. If we look at the journals *Cognitive Science* and *Artificial Intelligence*, a very considerable overlap is apparent: both are concerned with such topics as planning, errors, problem solving, learning, memory, inference and how to represent knowledge. Very often the same workers contribute to both journals. Also, as with many interdisciplinary fields, there is no homogeneity of approach (Mehler and Franck, 1981). Consequently, any distinction I draw will have an element of arbitrariness but, put very crudely, cognitive scientists *study* behaviour while AI workers *model* it. Typical of the former would be studies of the *differences* between novices' and experts' behaviour – how they approach or describe tasks, systems, languages, facts, procedures, etc. One important methodology here is the collection of 'protocols', or detailed records, of what subjects do (e.g. Ericsson and Simon, 1980). An example of this would be asking someone to write an essay, while you record their spoken comments for later transcription (with all the 'um's', pauses and expletives undeleted); the essay transcript and the writer's notes constitute a sufficiently detailed record to count as a protocol. (I give further practical examples of the cognitive science approach later, but stay at an abstract level for the moment.) Cognitive scientists also try to *discover* or formulate principles that orient the acquisition of expertise, such as the capability of experts to form mental models of how a system functions, given that system's constituents and interactions, or the capability of experts to write fluently. In similarly abstract terms, AI workers would investigate the simulation of these behaviours using computer programs to represent or model individuals' activities: making machines do things that would require intelligence if done by men.

Those for whom the term 'artificial intelligence' conjures up an image of a computer with which one can converse and reason in a manner indistinguishable from a human, may not feel reassured by the dialogue below, which its author gave as an example of a possible interaction with an *intelligent* information retrieval system (Wiseman, 1981:133):

I want to plan by summer holiday.

ABROAD?

Yes . . . try Sasskatoon.

DO YOU WANT FLIGHT DETAILS?

Not yet. Summer temperatures.

CLIMATE IS . . .

And where shall I stay?

HOTEL LIST IS . . .

However, I must emphasise that such a system does not yet exist, nor is our knowledge of human communication detailed enough to implement it in the foreseeable future, leaving aside its desirability. A brief illustration of the complexities involved is provided by studies in which two people, in different rooms, solved problems by conversing with each other via a range of media including handwriting, typing, voice and 'communication rich' (face-to-face). This revealed how the content, type and style of dialogue (e.g. its ambiguity, informality and redundancy) vary and depend on the task and the context (Chapanis, 1980). As we see in later sections, similar complexities come to light when we study other information-related activities, ranging from those so common as to be normally little thought about (e.g. writing, reading to study and reading as browsing) to those which are more vocational (e.g. editing, purposeful information retrieval).

Towards cognitive engineering

As we saw in the last section, cognitive science and AI workers are concerned with such matters as studying error-free and error-full performance. This involves both collecting information on the details of that performance and thinking about what kinds of processing mechanisms are involved – perhaps with a view to modelling how we perform the task under investigation. However, many tasks are too complex to be studied in this way, so early researchers partitioned them into more manageable and somewhat artificial sub-tasks. Just as we sometimes cannot see the wood for the trees, this reductionist approach often revealed a lot about the details of the sub-tasks but did not clarify how they interacted. Models built using those research findings were patently unrealistic, dealing with tasks '. . . in a sensible fashion, one at a time, rather than in the inelegant, cluttered human fashion of attempting to think of everything at once . . . with its virtue of creativity and flexibility' (Norman, 1981:1098). However, significant numbers of researchers are now turning once again to more complex tasks, ranging from those considered here to the mundane and everyday (e.g. using arithmetic in shopping).

Typical of the trend towards more practical research is a paper provocatively entitled 'Are we ready for a cognitive engineering?'. In this, Card and Newell (1981) observe the irony that while psychology is often criticised for its academic concern with fine details of the determinants of psychological performance, it is our lack of understanding of enough of those fine details that sets limits on the usability of computer-based equipment. Also, such information as psychology has to offer is often unusable by workers with other specialities. Card and Newell argue that a re-appraisal is necessary. What we need is:

> a psychology useful for engineering design . . . [which gives us] the ability to do *task analysis* (determining the specific, rational means of accomplishing various goals), *calculation* [predictions of behaviour] . . . and *approximation* (simplification of the task and of psychological theory)' (emphasis added).

To illustrate what can be achieved with current knowledge, they then derive a single model – the 'Model Human Processor' – which includes numerous results from cognitive psychology. Their paper elaborates on the details and shows how the model successfully predicts the time different people take on three sample tasks: morse code listening rate; reaching to a button; and reaction time. As they say, while there are many improvements that can be made to their model, a cognitive engineering is becoming feasible in certain well structured areas, of which copy editing and typing would be good examples.

Models of typing

Like the man who discovered that all his life he had been talking in prose, typists may be amazed to learn that the single task of typing requires multiple control of 60 tendons and 30 joints, just to move the fingers. Further, skilled typing is a classic example of a cooperative (rather than competitive) action by parts of the body:

> finger movements start several letters ahead of their scheduled arrival time, oftentimes out of sequence of the final temporal order in which they are made. It is as if each finger starts as soon as it can towards its intended target, and the hand appears to cooperate, configuring itself towards as many targets at a time as possible. (Norman, 1981:1099)

Knowing more about this is important because other skills involving high speed performance may develop similarly. Here, becoming expert involves a change from doing but a single acion at a time ('see-and-peck' typing) to touch-typing using both hands at once, of interest because it involves overlapping, cooperative performance of several simultaneous acts.

Some of the clues to how typists learn to touch-type may be provided by looking at the kinds of errors they make, such as typing 'bokk' rather than 'book'. This approach has been coupled with computer simulation of typing by Rumelhart and Norman (1981). Their computer model produces an output display of 'hands' and 'fingers' moving over a keyboard. It reproduces many of the major phenomena of typing, although it is not claimed that it offers a complete account of the conventional typing process, let alone the use of word-processors.

The keystroke-level model

If we wish to evaluate an interactive computer system such as a word-processor, we have to consider the range of possible users. For example, some users will be better typists than others; some will be casual users and others will use it a lot; and some will know a lot, others a little, about the tasks to be carried out on the system. We also have to bear in mind the software and hardware of the computer – the range of facilities available, the help offered users, the speed of response, etc. With so many variables, it is hardly surprising that the most fruitful evaluation studies to date have – as with typing – concentrated on expert users performing routine tasks.

In any study of performance we will need some kind of metric – some measure of that performance. Once we have decided on a representative task and a metric for that task, we need to collect detailed records, or 'protocols', of actual interactions between the user and the system (whether the 'system' is a piece of paper or a terminal). Both typewriters and word-processors are keyboard instruments so it is natural to think of recording users' key-depressions – 'keystroking' being the vogue term for this.

This prototypical model of user behaviour in this context is actually called the *keystroke-level* model of performance. The central idea behind the model is '. . . that the time for an expert to do a task on an interactive system is determined by the time it takes to do the keystrokes' (Card et al., 1980:397). Greatly simplified, to use this model we need to know the precise steps involved in the method adopted by a user to carry out the task; we then count the number of keystrokes required, and multiply by the time per keystroke to get the total time. More precisely, we also need information on: the 'command language' of the editing system, and the system response time. Given this information, we can predict the best performance obtainable with the system, that is, the error-free performance time of an expert user on the task. The significant thing about the model is that it makes explicit the significant operations required to accomplish the task being looked at, so allowing designers to *predict* and *quantify* actual user behaviour. Empirical validation of the model has been carried out on a variety of tasks, such as adding a box to a diagram or removing overlapping graphic elements using a graphics editor with or without visual equivalents – 'iconic representations' – of the commands. However, the task which received most attention initially was a simple form of copy editing – script editing, or the correction of a marked-up manuscript. The specific sub-tasks looked at ranged from word substitutions (simple) to moving sentences within paragraphs (more difficult). More demanding forms of editing by experts remain to be investigated using this model.

Electronic editing

Five types of editing are common on computer-based text-handling equipment (Thimbleby, 1981):

- composition (manipulating layout, experimenting with various ways of phrasing ideas);
- copy typing or copy editing (in which we are primarily typing new material or material from copy, with some occasional corrections);
- 'cut-and-paste' (in which we move sizeable portions of a document within the document);
- script editing (following written corrections to existing text); and
- spelling correction (manipulating single letters).

Using an electronic editing system rather than paper for editing may speed up many of these and certainly results in a more legible draft for the printer. Among the disadvantages of electronic editing are: access to the equipment and eyestrain (see also, Lefrere, 1981). Not least of the disadvantages is the need for special training. This training should take account of others' experience of learning to use computers. For example, beginning computer programmers often exhibit the 'working-set' syndrome: powerful constructs and techniques are not made part of their repertoire if it is possible for them to build up tortuous combinations of elementary ones. I think the editing analogue is learning how to carry out repeated changes, such as in correcting the style of references within a manuscript. Typically, the user of a word-processor will have been introduced to the elementary features of the machine before meeting the advanced features. Thus, 'insert', 'delete' and 'search-and-replace' may be encountered, used and learned before the user is trained to use 'iteration' commands or global combinations of elementary commands. Here, the optimal (fastest) method of changing the references might be to use an iteration command, but it is very common to find users who persist in deploying lower-level procedures, even though each reference has to be altered sequentially and in stages ('find', 'delete', 'insert' etc.).

Even if routine tasks can be carried out quickly, the system may not help with less routine types of editing. For example, consider the problem of 'manuscript surprises'. A manuscript surprise is a problem which is revealed only in careful editing:

after making what seems to be a reasonable decision, we encounter '. . . new information that makes it necessary to revise the previously edited elements in a linked series' (Farkas and Farkas, 1981:16). Those surprises may be the result of several types of author-generated inconsistencies, including:

– poor organisation of the ideas in the manuscript;
– mechanical inconsistencies (e.g. abbreviations; spelling; other aspects of 'house style'); and
– content inconsistencies (e.g. the use of technical terms).

In traditional, paper-based editing, each editor will have his own ways of ensuring consistency and keeping track of decisions made. If such 'housekeeping' can be delegated to electronic systems, this 'intelligence-augmentation' (Engelbart and English, 1968) could make it easier to undertake such tasks as obtaining an overview of the text; cancelling previous decisions; and seeing the effect of possible changes. These facilities are not yet available on stand-alone word-processors, although they have been available for some time on more expensive, mainframe computer systems.

Machines to help us to write?

Writing aids can be concerned with handwriting (calligraphy); spelling (orthography); and content and style (what to say and how to say it, without using unnecessarily long words!). Typewriters certainly help with the first of these; word-processors are said to help with the second; but what of the third? Many authors find an editor's help essential in making their manuscript intelligible. Only rarely does this 'editing for sense' make matters worse (e.g. Swaney et al., 1981). Yet some authors lack access to editors and others find the editor's help comes too soon after writing – they do not want their work appraised by others before they can 'distance' themselves emotionally from what they have written. While computer technology is not yet sufficiently advanced for computers to act as authors (Mann and Moore, 1980), or even to provide authors with rough drafts which they can polish, they may hope for some computer help somewhere in the writing process.

Consider the pro's and con's. It is true that, for those who can type quickly, being able to insert text or to 'cut and paste' can be easier using a screen than using conventional, paper-based methods. (But inventions often embody aspects of the things they replace; think of how early cars looked like carriages. Current text-editing systems are programmed to imitate several undesirable features of paper, according to Goldstein (1981), so new designs are necessary if we are to know whether qualitative improvements in people's ability to write are possible.) It is also true that microcomputer programs exist which purport to help with certain aspects of writing, such as correcting our spelling. (In reality, 'spelling correction' programs cannot spot contextual errors such as typing 'has' rather than 'had' – easy to do on a typewriter.) Whatever the reality, people believe what they want. Thus it is often asserted that using a word-processor rather than paper will help us to write. Despite writing this on a word-processor, I would argue that being able to change your mind or leave the dictionary on the shelf are but tiny parts of being helped to write.

'Make this the year you learn to write!' proclaims one advertisement for a postal course in my Sunday paper. The tips provided may be useful to budding journalists, but it is clear that many aspects of writing are still poorly understood. This is an active research field, in which the workers recognise that there is no best way to understand how people write, and no aspect of writing that is more deserving of study than another. Yet one particular viewpoint, that of writing as *cognitive processing* (e.g. Hayes and Flower, 1981), seems both relevant to, and ignored by, designers of word-processors.

It is common knowledge that writers differ. For example, some plan their essays completely before writing a word, while others never seem to look ahead. Again, some seem never to think of their readers, while others continually worry about them. Also as with other activities such as studying, an individual may adhere to one particular way of writing or change according to circumstance. Characterising such variation is difficult, but is essential if we want to specify the processes involved in writing and accurately describe their sequence and interaction. In Hayes and Flowers' words (op. cit.:390):

> When writers construct sentences, we want to know how they handle such multiple constraints as the requirement for correct grammar, appropriate tone, accuracy of meaning, and smooth transition.

Acquiring such information and using it to construct a model of the writing process is actually quite involved and has occupied Hayes' and other groups for several years without an end coming in sight. However, they are already able to provide a fairly refined description of several writing styles, including:

- 'depth first' (polishing each sentence before considering the next);
- 'get it down as you think of it, then review';
- 'perfect first draft'; and
- 'breadth first' (not writing a sentence until the whole essay is planned).

Among their findings are two which may be of interest: that some writing processes (e.g. editing, generating) take precedence over others and may *interrupt* them in characteristic ways (e.g. in the middle of writing a sentence, to change a word; or at the end of a sentence, to link to the next); and that certain writing processes (e.g. generating and organising) are often *hierarchically organised.* The significance of the first finding, I think, is that the frequency with which people interrupt the physical act of writing to search for more appropriate phrases, indicates that a thesaurus-like facility could be useful to many authors. Similarly, the second, organisational or *planning* finding indicates to me that this facet of writing should be supported in some active way by text-handling systems. For example, it could be useful to store one's initial ideas (e.g. as a 'spray pattern') and refer to them while writing. Neither paper nor current commercial word-processors help very much here. As with paper, the author generally has to work under various constraints (e.g. length; numbering of captions; cross-referencing and citations) which make it cumbersome to explore alternatives and encourage premature commitment to a particular plan. Those constraints also discourage exploration of writing in non-traditional ways, which could help readers (e.g. Jewett, 1981).

If adults have difficulty writing, it could be that they were never encouraged to re-draft material when they were young. Being able to re-draft is a good first step in developing creative writing, yet children also need to appreciate language construction and to explore ways of transforming text to suit its audience and its function. There are several computer-based ways of helping with this. One approach is to use the computer to direct attention, ask questions and help keep track of how an essay is going (Woodruff et al., 1981). A variant on this, intended for adults, is to use the computer 'as a device to articulate a formal theory of argument structure', providing 'planning schemata to represent different arguments such as argument-by-induction, argument-by-authority, and argument-by-deductive proof . . . with slots for the various positions that an argument requires' (Goldstein, 1981:147). Whether we yet know enough about argumentation and its effects on readers is disputable, of course (Whalley, 1982). Another approach is to provide a number of tools which may be of use, as in one study where children used a word-processor/text-transformer, an automated thesaurus and dictionary, a spelling corrector, a story planner and a sentence generator (Sharples, 1981). Yet another possible approach is to alert children to stylistic deficiencies in their work, an approach which was originally developed for adults (Angier, 1981; Cherry, 1981). This last approach relies upon having programs to evaluate surface features of text such as readability, sentence and word length, sentence type, word usage and sentence openers. The associated copy-editing programs also alert authors or editors to possible overwordiness, bad diction, split infinitives, etc., and represent the current 'state-of-the-art' in commercially-available author aids, although the lack of understanding by the system of what it is processing limits its use to authors of technical papers.

The future

Without looking far into the future, machines are likely to be given a quasi-understanding of increasing numbers of the tasks they undertake for us. A simple but trivial example of the understanding which we take for granted is our ability to see differences between and meanings in the instructions we give to a typewriter when we type a letter or a memo. It is unrealistic to expect the next generation of machines to understand the details of what we write, but we can expect to find a way of *describing* (or rather, giving the machine an 'internal representation' of) gross features of the strings of characters and spaces in a limited range of document types, so that it can 'recognise' parts of a letter such as the address and the opening and closing salutations.

A more interesting example is the task of checking references. At the moment, this is a time-consuming task. To speed this up, an author or editor may ask for an on-line search to be carried out on an appropriate database. Microcomputers already help a little here, in that we can use them to store and recall frequently used sequences of keystrokes, essential for connection to the database and searching through it (e.g. Williams, 1980). Yet much of AI research is concerned with pattern recognition. Preliminary work at the Open University indicates it is quite likely that machines could be taught to recognise parts of citations and references (e.g. type of source; date; edition; author and editor). The next stage would be to teach the machines to use this knowledge to identify gaps in those references and to construct and carry out consistency checks (of your citations – or even of the accuracy of the database) for you. Two other aspects of information retrieval which may be affected by AI techniques are: how to improve retrieval; and how to adapt to users' changing needs and purposes. One possible way to improve retrieval is to use the information contained in documents to hand, if they are in machine-readable form. The text of such a document often contains qualifying statements about the documents it itself cites, but teaching a computer to recognise and use those 'citing statements' (O'Connor, 1980) is not a trivial task. Similarly complex is the task of creating a local database which adapts to its users' needs. If we take an AI approach to this, the task may be tractable, as indicated by preliminary work on designing a system which learns by making inferences about the success of past searches and adapts accordingly (Holland, 1980).

Turning to other aspects of documentation such as writing, editing and design, our increasing appreciation of the details of these activities makes it easier to codify the knowledge-base being drawn upon. One consequence is that it becomes possible for one person to combine these roles if necessary (Macdonald-Ross and Waller, 1976). Also, if the relevant expertise can be expressed in fairly precise terms, we can expect to see the emergence of 'expert systems' (Lefrere, Waller and Whalley, 1980). These are computer systems able themselves to act as consultants, so the next edition of this volume may have quite a different set of authors!

References

Angier, N. (1981), 'Bell's Lettres', *Discover*, July, 78–9.

Bush, V. (1945), 'As we may think', *Atlantic Monthly*, July, 101–8.

Card, S. K., Moran, T. P. and Newell, A. (1980), 'The keystroke-level model for user performance time with interactive systems', *Communications of the ACM*, 23, 396–410.

Card, S. K. and Newell, A. (1981), 'Are we ready for a cognitive engineering?' in *Proceedings of the 3rd Annual Conference of the Cognitive Science Society*.

Chapanis, A. (1980), in B. Shackel (ed.), *Man/Computer Interaction*, Alphen aan den Rijn, Sijthof and Noordhoff.

Cherry, L. (1981), 'Computer aids for writers', *SIGPLAN Notices*, 16(6), 61–7.

Engelbart, D. C. and English, W. K. (1968), 'A research center for augmenting human intellect', in *AFIPS Conference Proceedings, 33*.

Ericsson, K. A. and Simon, H. A. (1980), 'Verbal reports as data', *Psychological Review*, 47(3), 215–51.

Farkas, D. K. and Farkas, N. (1981), 'Manuscript surprises: a problem in copy editing', *Technical Communication*, 28(2), 16–18.

Fox, M. S. and Palay, A. J. (1982), Machine-assisted browsing for the naive user, in J. L. Divilbiss (ed.), *Public Access to Library Automation*, Champaign, University of Illinois.

Goldstein, I. (1981), 'Writing with a computer', in *Proceedings of the 3rd Annual Conference of the Cognitive Science Society*.

Hayes, J. R. and Flower, L. S. (1980), 'Writing as problem solving', *Visible Language*, 14, 388–99.

Heyel, C. (1969), *Computers, Office Machines, and the New Information Technology*, London, Macmillan.

Holland, J. H. (1980), 'Adaptive algorithms for discovering and using general patterns in growing knowledge bases', *Int. J. of Policy Analysis and Information Systems*, 4, 245–68.

Jewett, D. L. (1981), 'Multi-level writing in theory and practice', *Visible Language*, 15, 32–40.

Lefrere, P. (1981), 'Editors' roles in on-line journals', *J. of Research Communication Studies*, 3, 157–67.

Lefrere, P. and Whalley, P. (1979), 'Computer assistance in multimedia educational publishing', in *Proceedings of the PIRA/RPS International Conference on Trends in Educational Publishing.*

Lefrere, P., Waller, R. and Whalley, P. (1980), 'Expert systems in educational technology? in R. Winterburn and L. Evans, (eds), *Aspects of Educational Technology 14*, London, Kogan Page.

Macdonald-Ross, M. and Waller, R. (1976), 'The transformer', *Penrose Annual*, 69, 141–52.

Mann, W. C. and Moore, J. A. (1980), *Computer as Author – Results and Prospects.* Report ISI/RR-79-82, University of Southern California.

McCullogh, W. S. (1965), *Embodiments of Mind*, Cambridge, Mass., MIT Press.

Mehler, J. and Franck, S. (1981), 'Editorial', *Cognition*, 10, 1–5.

Moran, T. P. (1981), 'An applied psychology of the user', *Computing Surveys*, 13, 1–11.

Norman, D. A. (1981), A psychologist views human processing: human errors and other phenomena suggest processing mechanisms', in *Proceedings of the 7th International Joint Conference on Artificial Intelligence.*

Robertson, G., McCracken, D. and Newell, A. (1981), 'The ZOG approach to man–machine communication', *Int. J. Man–machine Studies*, 14, 461–88.

Rumelhart, D. E. and Norman, D. A. (1981), *Simulating a Skilled Typist: A Study of Skilled Cognitive-motor Performance.* Technical Report 102, Center for Human Information Processing, University of California, San Diego.

Senders, J. W. (1980), 'The Electronic Journal', in L. J. Anthony (ed.) Eurim 4. A European Conference on Innovation in Primary Publication: impact on producers and users. London, Aslib.

Sharples, M. (1981), 'Microcomputers and creative writing', in J. A. M. Howe and P. M. Ross (eds), *Microcomputers in Secondary Education: Issues and Techniques*, London, Kogan Page.

Strawhorn, J. M. (1980), 'Future methods and techniques', in P. Hills (ed.), *The Future of the Printed Word*, London, Frances Pinter.

Swaney, J. H., Janik, C. J., Bond, S. J. and Hayes, J. R. (1981), 'Editing for comprehension: improving the process through reading protocols', *Document Design Project Report 14*, Carnegie-Mellon University.

Thimbleby, H. W. (1981), A word boundary algorithm for text processing', *The Computer Journal*, 24, 249–55.

Whalley, P. (1982), 'Argument in text and the reading process' in A.

Flammer and W. Kintsch, (eds), *Discourse Processing*, Amsterdam, North-Holland.

Williams, P. W. (1980), 'Intelligent access to on-line systems', in *Proceedings of the 4th International On-line Information Meeting*, Oxford, Learned Information.

Wiseman, N. (1981), 'Improvements to Television', *Information Design Journal*, 2, 131–135.

Woodruff, E., Bereiter, C., and Scardamalia, M. (1981), 'Experiments in computer-assisted composition', presented at AERA Conference, Los Angeles.

10. LIBRARIES AND THE NEW TECHNOLOGY: A SELECT BIBLIOGRAPHY

GEOFFREY E. HAMILTON
Librarian of the Library Association
The British Library,
Library Association Library,
London.

This list provides references to a selection from the large number of articles, books, conference papers and reports which discuss new technology in relation to libraries and information units. The three main divisions of this list cover library implications of new technology, with a few items which treat particular types of library; studies of particular technologies which deal specifically with library applications; and some forecasts of what may happen to libraries and librarians in the future. All items listed are held by the Library Association Library. The original list was issued November 1980. It has been supplemented in December 1981 by a further listing comprising chiefly items published since the first list was compiled. The plan of arrangement is similar to the earlier list, and the numbering of items follows on from that list. The contents list below covers both the original list and the supplement.

CONTENTS — Item numbers

Implications of the new technology

General

1. Avram, Henriette D., The impact of technology on legislation affecting libraries. *Journal of Library Automation* 12 (4) Dec. 1979, 355-61. (Also in *IFLA Journal*, 6 (1) Feb. 1980, 8-12.)

   Reviews problems caused by introduction of new technology in the context of national legislation. Suggests that libraries could be seriously hampered by legislation regulating free flow of information or imposing tariffs.

2. Barrentine, James K., The future of computer technology in library networking. In *Networks for Networkers: Critical Issues in Cooperative Library Development*. Mansell, 1980, 136-53.

   Sees a future in which the technology is used in a national network which will provide a communications facility between terminals in libraries and any supplier of information.

3. Great Britain. *Parliament. House of Commons. Education, Science and Arts Committee*. Fourth report from the . . . Committee, session 1979-80: *Information Storage and Retrieval in the British Library Service*; together with the minutes of evidence and appendices. HMSO, 1980, xxix, 158p. (HC 767, session 1979-80).

   The current impact of new technology is among the topics covered in the report and evidence. Memoranda printed include one on the New Technology Group of the British Library, by Peter T. Kirstein (Appendix 2, pp. 97–8).

4. *The Information Society: Issues and Answers*. American Library Association's Presidential Commission for the 1977 Detroit Annual Conference; edited by E. J. Josey. Mansell, 1978, xxii, 133p.
   Contents include papers on the impact on libraries of technology, and of social and economic change, and on the new role of the librarian in the information age, and literature reviews for each of those four topics.

5. Lancaster, F. W., *Libraries and the Information Age*. In the ALA yearbook 1980. Chicago: American Library Association, 1980, 9–19.
   Predicts the role of libraries and librarians, future trends in special and public libraries and the growth of private information practices.

6. McGraw, Harold W., Responding to information needs in the 1980s. *Wilson Library Bulletin*, 54 (3) Nov. 1979, 160–4.
   Reviews library complications of the new technology, with comments on US action in this field.

7. Mathews, William D. Advances in electronic technologies. *Journal of Library Automation*, 11 (4) Dec. 1978, 299–307.
   Considers electronic hardware for processing, storing and transmitting information.

8. Martin, Susan K. *Technology in Libraries, 1960–2000*. Arlington, VA: Educational Resources Information Center, 1978. 15 pp. (ERIC Report ED-158 764).

9. Nielsen, Brian, Online bibliographic searching and the deprofessionalisation of librarianship. *Online Review*, 4 (3) Sept. 1980, 215–24.
   Argues that trends in the development of technology suggest that the status-enhancement conferred upon

librarianship by online bibliographic searching may be short-lived.

10. 1985: new technology for libraries. *Library Journal*, 105 (13) July 1980, 1473-8.
    Forecasts by five executives of information companies – Roger Summit, Eugene Garfield, Isaac Auerbach, Janet Egeland, Walter Bauer – of which product or technical innovation promises most for information services from libraries, and of what libraries will be like in 1985.

11. Popper, P., On-line information scenario for the eighties and its social implications. In *Information policy for the 1980s:Proceedings of the Eusidic Conference . . . October 1978.* Oxford: Learned Information, 1979, 47–52.
    Presents a scenario for a paperless society, and poses questions relating to management of the new technology, information overload and economic aspects.

12. Shaughnessy, Thomas W., Redesigning library jobs. *Journal of the American Society for Information Science*, 29 (4) July 1978, 187–90.
    Examines the impact of rapid technological change on library jobs.

13. Summit, Roger K., Problems and challenges in the information society – the next twenty years. *Journal of Information Science*, 1 (4) Oct. 1979, 223–6.
    Predicts that the 1980s' and 90s' problems will be administrative and political, rather than technical as was the case in the 1960s and 70s.

14. Technology and libraries. *Library Association Record*, 81 (10) Oct. 1979, 479.
    Proceedings of a debate at the LA's 1979 AGM on a proposal to set up a working party to look at the implications of the new technology for libraries.

15. Turoff, Murray and Spector, Marian, Libraries and the implications of computer technology. In *AFIPS Conference*

Proceedings, vol. 45, 1976 National Computer Conference. Montvale, NJ: AFIPS Press, 1976, 701-8.
As well as examining the impact of computer and information technology on the role and mission of the library, considers the impact of the library on the technology. Examines the library as a mechanism for making computer and information technology directly available to the public.

16. Usherwood, R. C., Socio-economic implications of the new information technology. *Aslib Proceedings*, 32 (6) June 1980, 276-8.
Attacks the 'specious argument' that the cost of the new technology justifies charges for public library services.

17. Williams, P. W., New opportunities from information technology. *Journal of Information Science*, 2 (1) Aug. 1980, 29-36.
Surveys opportunities for the smaller information unit, and reviews the costs and difficulties of the possible options.

Medical libraries

18. Valentine, Paul, Computer the central home. *Library Association Record*, 82 (2) Feb. 1980, 65, 67, 69.
Discusses the impact of technology on medical libraries.

Public libraries

See also items 32, 33.

19. Borgman, Christine L. and Korfhage, Robert R., The public library interface to personal computer systems. In *The Information Age in Perspective: Proceedings of the Annual Meeting of the American Society for Information Science*, vol. 15. White Plains, NY: Knowledge Industry Publications, 1978, 41-3.
Examines the role that the American public library ought to play in the provision of information when microprocessors have given the public a computing capability.

20. Huse, Roy, Libraries: automation the key to development. *Municipal Journal* 88 (3) 18 Jan. 1980, 82–3.
Examines likely developments in UK public libraries in the 1980s, suggesting that the way to reconcile increasing demand, an economic crisis, a static growth rate and cuts in public expenditure is through automation.

Special libraries

21. Patten, M. N., The special librarian of the future. In K. C. Harrison (ed.), *Prospects for British Librarianship*, Library Association, 1976, 230–44.

Studies of particular technologies

Facsimile transmission

22. Engelke, Hans, Telefacsimile use in US libraries. *Interlending Review*, 6 (2) April, 1978, 44–9.

23. Saffady, William, Facsimile transmission for libraries: technology and application design. *Library Technology Reports*, 14 (5) Sept/Oct. 1978, 445–531.

Microforms

24. Williams, B. J. S., Microforms as an information medium: a review of developments for librarians and publishers. *Reprographics Quarterly*, 12 (2) Spring 1979, 50–3. 19 refs.
The review covers computer input and output microfilm (CIM and COM), image quality and reading equipment and general micropublishing.

Microprocessors

25. *Microprocessors and Intelligence*, Proceedings of an Aslib seminar held 14–15 May 1979 at the Holiday Inn, Slough, edited by L. J. Anthony. Aslib, 1980. vii, 62pp.
Includes texts of five papers, amongst which are 'Developments in raw technology' by C. J. B. Hawkins; 'Word processing' by K. Edwards; 'The effect of microtechnology on information management' by P. W. Williams.

26. Williams, P. W., The potential of the microprocessor in library and information work. *Aslib Proceedings*, 31 (4) April 1979, 202–9.

Includes recommendations on tasks which are most suitable for the use of microprocessors.

27. Winfield, R. P., An informal survey of operational microprocessor-based systems, Autumn 1979. *Program*, 14 (3) July 1980, 121–9.

Describes ten systems in use in UK libraries and information units.

Telecommunications

28. Changing communications systems and the future of libraries: a panel topic at ALA Annual Conference Dallas, 1979. *Advanced Technology Libraries*, 8 (8) Aug. 1979, 1–4.

Communications experts comment on the library implications of electronic message systems, home video display terminals and communications satellites.

29. Isotta, N. E. C., The future impact of telecommunications on information science. *Journal of Information Science*, 1 (5) Jan. 1980, 249–58.

Postulates an ideal information system, which is compared with present technology and with what will probably be available within the next ten years.

Video

30. Bahr, Alice Harrison, *Video in Libraries: A Status Report 1979–80.* White Plains, NY: Knowledge Industry Publications, 1980. 119pp.

Chapter 6, 'The future: new technologies' (pp. 87–94)

Videotex

See also item 28.

31. Carr, Reg, Prestel in the test trial: an academic library user looks back. *Journal of Librarianship*, 12 (3) July 1980, 145–58.

Discusses the philosophy and design criteria behind Prestel, reviews trial usage at the University of Aston, and highlights advantages and disadvantages of Prestel as an information source in academic libraries.

32. Cherry, Susan Spaeth, The new TV information systems. *American Libraries*, 11 (2) Feb. 1980, 94–8, 108–10.
    Review of videotext and relevant systems throughout the world and their likely impact on the public library.

33. Martyn, John, Prestel and public libraries: an LA/Aslib experiment. *Aslib Proceedings*, 31 (5) May 1979, 216–36.
    Comments on the future implications for libraries of a community information service on Prestel with a database created by public library systems.

Word processing

See also item 25.

34. Saffady, William, The use of word processing equipment in libraries. *Library Technology Reports*, 13 (5) Sept. 1977, 487-91.

35. Whitehead, J. B., Developments in word processing systems and their application to information needs. *Aslib Proceedings*, 32 (3), March 1980, 118–33.

Libraries in the future

36. Lancaster, F. W., Drasgow, Laura S. and Marks, Ellen B. The changing face of the library: a look at libraries and librarians in the year 2001. *Collection Management*, 3 (1) Spring 1979, 55–77.
    Draft of a scenario to show the likely role of the library in a paperless society.

37. Lancaster, F. W., Whither libraries? or, Wither libraries. *College & Research Libraries*, 39 (5) Sept. 1978, 345–57.
    Presents a scenario for a paperless communication and

describes some technological achievements that lend credibility to the scenario.

38. Lancaster, F. W., *Toward Paperless Information Systems*. New York: Academic Press, 1978. xii, 179pp.
    Includes a discussion of the role of the library in a paperless society (pp. 153–9).

39. Lewis, D. A. Today's challenge – tomorrow's choice: change or be changed, or the doomsday scenario, Mk. 2. Welwyn Garden City: the author (c/o ICI Plastics Division), 1980. 34pp.
    Paper presented at the Institute of Information Scientists Annual Conference 1980. Includes a review of technological trends and concludes that they, and other trends, indicate the demise of the information scientist and the librarian by the year 2000.

40. Licklider, J. C. R. *Libraries of the Future*. Cambridge, MA: MIT Press, 1965. xvii, 219pp.
    Summary report of a study sponsored by the Council on Library Resources. The 'future' was defined as the year 2000.

41. Meadow, Charles T., Information science and scientists in 2001. *Journal of Information Science*, 1 (4) Oct. 1979, 217–22.
    Discusses how technological innovations will lead to changes in the way information is acquired, transmitted and used.

42. Musmann, Klaus, Will there be a role for librarians and libraries in the post-industrial society? *Libri*, 28 (3) 1978, 228–34.
    Reaches an optimistic conclusion.

43. *The Role of the Library in an Electronic Society*, edited by F. Wilfred Lancaster. Urbana-Champaign: University of Illinois Graduate School of Library Science, 1980. 200pp.
    Proceedings of the 1979 Clinic on Library Applications

of Data Processing. Topics covered include: 'Electronic information exchange and its impact on libraries' (117–34); 'A forecast for the future of computer technology' (135–61); 'A scenario on the role of the library in an electronic society' (162–89).

44. Salton, Gerald, Proposals for a dynamic library. In Leigh Estabrook (ed.) *Libraries in Post-Industrial Society*, Mansell, 1977, 272–305.
   'A blueprint for a new mechanized library environment.'

45. Strozik, Teresa, Lowen, Walter and Wakefield, Rowan A. *Input, Output, Throughput and Kaput: Position Paper on Technology.* Arlington, VA: Educational Resources Information Center, 1978. 23pp. (ERIC report ED-163 899)
   A review of the history of libraries from the viewpoint of the year 2080.

Implications of new technology (Supplement)

General

see also item 66

46. Barron, Ian and Curnow, Ray, *The Future with Microelectronics: Forecasting the Effects of Information Technology.* Frances Pinter, 1979. 243pp.
   Not specifically concerned with libraries, but makes some interesting points about the information revolution and briefly discusses uses of i.t. in connection with information services.

47. *Automation and Serials: Proceedings of the UK Serials Group Conference 1980*, edited by Margaret E. Graham. UK Serials Group, 1981.
   Includes a session on new technology, with papers on agency services, new technology in local systems, high technology in relation to library systems, and low budget approaches to new technology (pages 91–105).

48. Black, John B., New information technologies: some

observations on what is in store for libraries. *Inspel*, 15 (3) 1981, 145–53.

Summarises the primary implications of the new technologies for libraries under the headings of services, networks, staff, collections and materials, buildings and budgets.

49. Doll, Russell, Information technology and its socioeconomic and academic impact. *Online Review*, 5 (1) Feb. 1981, 37–46.

Suggests that information technology may bring about increases in power for those able to make best use of it, and the supplanting of nation states by multinational corporations. Unplanned changes arising from i.t. may affect research directions, library training, accessibility of source materials and the formation of library elites.

50. Featheringham, T. R. Paperless publishing and potential institutional change. *Scholarly Publishing*, 13 (1) Oct. 1981, 19–30.

Includes an assessment of potential impacts on libraries and librarians. Points out that librarians and information scientists have an obligation to ensure that automation does not turn into dehumanisation.

51. *The Future of the Printed Word: The Impact and the Implications of the New Communications Technology*, edited by Philip Hills. Frances Pinter, 1980. 172pp.

A book of readings by a group of specialists in the fields of publishing, librarianship, information science, computing and education. Contents include: 'Designing for the new communications technology' by Linda Reynolds, 81–98; 'Electronic alternatives to paper-based publishing in science and technology' by Donald W. King, 99–110.

52. Line, Maurice B., The paperless society — do we want it? Do we need it? *Technicalities*, 1 (12) Nov. 1981, 2–3.

Technology should extend the range of media available, not reduce it; and the use of technology should be determined by human needs, and not human behaviour by technology.'

53. Smith, Joan M., Simpson, Ian S. and Kemp, D. Alasdair *Report on Workshop on New Technology and Library/ Information Science Education, Newcastle upon Tyne Polytechnic, April 13-16, 1981.* School of Librarianship and Information Studies, Newcastle upon Tyne Polytechnic, 1981. 24pp. (British Library Research and Development Report no. 5668).

Concentrates on developments in microcomputers and videotex, with emphasis on the place of new technology in curriculum development.

54. Tedd, Lucy A., *The Teaching of Online Cataloguing and Searching and the Use of New Technology in UK Schools of Librarianship and Information Science.* British Library, 1981. 120pp. (British Library Research and Development Report No. 5616)

Chapter 3 describes use of microcomputers, intelligent terminals, viewdata systems etc. in UK library schools in late 1979.

55. Weil, Ben H. Technical-communication fundamentals in an era of technological change. *Journal of Chemical Information and Computer Sciences*, 21 (3) Aug. 1981, 119-22.

The quality of messages sent via high technology communication systems and the ease with which these can be read will be considerably lessened unless system designers apply more of what is known about textual presentation and unless authors pay better attention to how they write.

56. Wight, Tony, *A Discourse on Issues: an Explanatory Study of the Implications of Information Technology for UK Library and Information Work Manpower Planning.* Aslib, 1980. 190pp. (British Library Research and Development Report no. 5656).

Describes the pilot study of a proposed project to forecast and assess the impacts of information technology over the next twenty years. Among issues discussed are the role of libraries and information services; the prospects for microelectronics, telecommunications; office technology and applied artificial intelligence.

Public libraries

See items 64, 76, 77.

Special libraries

57. Information technology and special libraries. *Special libraries*, 72 (2) April 1981, 97–174.
    Contents include: 'Socio-political impact of information technology' by Marilyn K. Gell, 97–102; 'Telecommunications and facsimile' by Henry Voos, 118–21; 'The impact of office automation on libraries' by Robert M. Landau, 122–6; 'Information technology and personal responsibility' by Irving M. Klempner, 157–62; 'Information technology; a bibliography' by William F. Wright and Donald T. Hawkins, 163–74 (86 annotated entries).

Studies of particular technologies

Facsimile transmission

See items 57, 66

Holography

58. Voos, Henry, Implications of holography for information systems. *Journal of the American Society for Information Science*, 31 (6) Nov. 1980, 449–51.
    Describes the advantages of using holography for information storage and other applications.

Microprocessors

See also items 53, 54

59. Armstrong, C. J., Guy, R. F. and Painter J. D., Microcomputers or word processors in the library? *Reprographics Quarterly*, 14 (3) Summer 1981, 98–103.

Describes the characteristics, advantages and disadvantages of these two types of equipment, and discusses their library applications.

60. Burton, Paul F. The microcomputer in the smaller library. *SLA News*, (160) Nov–Dec. 1980, 175–8.
    Describes the Marindex system at Leith Nautical College Library.

61. Krueger, Donald R., Issues and applications of microcomputers for libraries. *Canadian Library Journal*, 38 (5) Oct. 1981, 281–6.

62. Lundeen, Gerald, The role of microcomputers in libraries. *Wilson Library Bulletin*, 55 (3) Nov. 1980, 178–85.
    Examines actual and potential applications of microcomputer systems in libraries, for technical and public services.

63. *Minis, Micros and Terminals for Libraries and Information Services*, edited by Alan Gilchrist. Heyden, 1981. 121pp. (British Computer Society Workshop Series, 1)
    Proceedings of a conference held at the National Computing Centre, Manchester, 6–7 Nov. 1980. 11 papers, including 'Information Technology – problems and opportunities by P. W. Williams (pp. 1–11); 'Intelligent terminals for library and information work' by P. L. Noerr (pp. 55–70); 'Application of a communicating word processor in an information department' by W. M. Henry (pp. 96–104).

64. Romans, Anne F. and Ransom, Stanley, A., An Apple a day: microcomputers in the public library. *American Libraries*, 11 (11) Dec. 1980, 691–3.
    Describes use of a microcomputer installed in the children's library in a rural public library system in New York State.

54. Vickery, A. and Brooks, H. Microcomputer, liberator or enslaver? In *4th International Online Information Meeting*,

London, 9–11 December 1980. Learned Information, 1980. pp. 387–96.

Assesses the role of microcomputers in library and information science with reference to the experience of the University of London Central Information Service.

Telecommunications

See also items 56, 57

66. *Telecommunications and Libraries: A Primer for Librarians and Information Managers*. White Plains, NY: Knowledge Industry, 1981. 187pp.

A collection of essays describing various kinds of new technology, including videotext, facsimile and optical disc technology. More general papers on 'Libraries and the transfer of information' by F. W. Lancaster, and 'Roadblocks to future ideal information transfer' by D. W. King are also included.

67. Veenstra, Robert J., Electronic mail has a future in the library. *Special Libraries*, 72 (4) Oct. 1981, 338–46.

'New technology and the promise of regulatory change [in the US] have created a potential for the instantaneous transmission of written messages and document delivery systems for interlibrary loans and other services anywhere in the world'.

Video

See also item 66

68. Barrett, R., *Developments in optical disc technology and the implications for information storage and retrieval*. British Library, 1981. 72pp. (British Library Research and Development Report no. 5623)

69. Boyle, Deirdre, Video fever. *Library Journal*, 106 (8) April 15 1981, 849–52.

Quotes the case of MIT's Slideathon (54,000 art and architecture slides accessible on a single laser videodisc) to cast doubt on the accuracy of the statement that 12,000 books could be stored on a single disc selling for $20.

70. Horder, Alan, *Videodiscs: their application to information storage and retrieval.* Hatfield: National Reprographic Centre for Documentation, 1979. 38pp. (NRCD publication no. 12)

71. Leclerq, Angie, Videodisc technology: equipment, software and educational applications. *Library Technology Reports,* 17 (4) July/Aug. 1981, 293-324.
    Discusses questions relating to videodisc systems which are of concern to librarians and suggests how a library should decide whether to spend some of its limited a/v resources on videodisc technology.

72. Singleton, Alan, The electronic journal and its relatives. *Scholarly Publishing*, 13 (1) Oct. 1981, 3-18.
    Includes section on 'Routes to the electronic journal', commenting on European developments in electronic document delivery including a proposed full-text store of journal articles on optical videodisc which could be used by the British Library Lending Division to satisfy loan requests.

73. Sneed, Charles, The videodisc revolution: what's ahead for libraries? *Wilson Library Bulletin*, 55 (3) Nov. 1980, 186-9.
    Describes existing systems and possible future developments.

74. *Video Involvement for Libraries: A Current Awareness Package for Professionals*, edited by Susan Spaeth Cherry. Chicago: American Library Association, 1981. 83pp.
    A compilation of articles on video that appeared in *American Libraries* between April 1979 and October 1980.

Videotex

See also items 53, 54, 66

75. Plakias, Mark, New electronic media: the future and co-operation. *RQ*, 20 (1) Fall 1980, 66–9.
'A discussion of the growing area of information delivery systems for the home and their impact on libraries.'

76. Sullivan, Catherine, *Impact of Prestel on public library branch services.* Aslib, 1981. 58pp.
Investigates the value and use of Prestel as a reference tool in public branch libraries.

77. Sullivan, Catherine and Oliver, David, *The impact of Prestel on public library reference activities.* Aslib, 1981. 99pp. (British Library Research and Development Report no. 5654)
The first major study of the application of Prestel in libraries for public access, investigating its use in six central reference libraries.

78. *Viewdata and Videotext 1980–81: A Worldwide Report* (Transcript of Viewdata '80, first world conference on viewdata, videotex and teletext). White Plains, NY: Knowledge Industry, 1980. 623pp.
Papers cover videotex activities in Europe, North America and Japan, technical and marketing aspects and videotex applications including electronic publishing.

Word processing

See also items 59, 63

79. Clough, R., New technology – new opportunity. In *4th International Online Information Meeting, London, 9–11 December 1980.* Learned Information, 1980. pp. 245–53.
Discusses word processing with particular reference to its potential in libraries and a library system based on the IBM ASSASSIN information retrieval package.

80. Moulton, Lynda W., Word processing equipment for information centers. *Special Libraries*, 71 (11) Nov. 1980, 492–6.

Describes use of a DEC WS-81 word processor by the Energy Economics Group of Arthur D. Little Inc. for a specialised information centre and library. Lists advantages and disadvantages encountered in using this equipment.

81. Whitehead, John, Word processing: an introduction and appraisal. *Journal of Documentation*, 36 (4) Dec. 1980, 313–41.
    A 'Progress in documentation' article, with a bibliography of 40 items.

82. Whitehead, John, Word processing and information management. *Aslib Proceedings*, 33 (9) Sept. 1981, 325–42.
    Includes section on applications in library and information functions.

Libraries in the future

See also item 56

83. Neill, S. D., Canadian libraries in 2010. Vancouver: Parabola, 1980. 144 pp.
    Reviewed in *Library Quarterly*, 51 (4) Oct. 1981, pp. 445–6 by Samuel Rothstein, who explains the book's very unusual structure and subject matter. Its scope is wider than the title implies.

84. Wooster, Harold, The bibliotaphic libraries of the year 2000. *Library Resources and Technical Services*, 25 (1) Jan/March 1981, 104–9.
    'This paper is an unsavoury mixture of library automation literature and science fiction.' The author embeds the ultimate amiable user interface in a framework of underground libraries connected by broadband communication networks and bicycle messengers.

11. NEW TECHNOLOGY AND INFORMATION TRANSFER: A SURVEY OF OPINIONS

PHILIP J. HILLS
Primary Communications Research Centre,
University of Leicester

At the 1981 Frankfurt Book Fair I talked to publishers interested in developments in the new technology and asked a good number of them to complete a questionnaire for me. The results of these and my discussions are given below. This is not intended to be a statistically representative survey of opinion but simply a guide to how some publishers with experience in this field are thinking about new developments in publishing and information transfer. My thanks to all those who contributed their opinions and my apologies to all those who, after reading this account, would have liked to contribute but did not get the opportunity.

A definition of new technology

There are many differing definitions of new technology generally covering the use of microcomputers, microprocessors, databases, information networks, etc. In this survey I asked publishers to consider all of these together with any new developments or equipment which might help the author, editor, publisher, printer, distributor and reader, i.e. all those involved in the process of information transfer.

How will the use of new technology develop in publishing?

It was felt that, initially at least, the greatest developments would occur with new technology helping the conventional publishing process and that entirely electronically based information networks would develop slowly.

There was general agreement that there will be an increasing link between the author, editor, publisher and printer. There would be more direct author/editor input by microbased technology often linked by networks or via satellites to publisher host computers. Texts would be stored on magnetic discs with consequent ease of editing, correction, etc.

Developments in high quality typesetting and laser printing controlled by computer would continue.

There was considerable realisation of the possibilities in the technical 'massaging' of textual material once captured electronically and it was felt that there would be an increasing use of screen generated graphics.

What type of information is best suited to developments in new technology?

A distinction was made between topical, ephemeral, rapidly changing and leisure or academic reading.

Electronically based networks with access from the home using a video screen were seen to be one increasing way in which the first type of rapidly changing information could be stored and retrieved. However, in terms of present developments the economic viability of such systems was questioned and it was felt that the public may not be prepared to pay for the real costs of the information. Initially, therefore, mainly subsidised information from government or industrial organisations might be possible rather than the total concept of the electronic newspaper.

The possible interactive teleordering facilities of electronic systems were felt to have considerable possibilities. The present use of such systems for the acquisition of business information and their use for travel bookings etc., was obviously an indication of future uses where the rapid, accurate transfer of topical information and current data are required. The probable developments in our electronic 'cash-less' society were mentioned.

Two publishers put the point that although topical information could be transferred from place to place faster by electronic means they felt that people generally like to have a print on paper record so that they can easily refer to it.

It was felt that there would be a growth of on-demand

publishing and that the computer was especially useful for generating encyclopedia-like publications.

How might the work of those involved in the process of information transfer change?

Again the distinction between types of information was mentioned and it was felt that as topical, ephemeral and rapidly changing information moved towards dissemination by electronic means so the work of authors, journalists, publishers and printers would be affected. Those people handling information 'of a more permanent kind' were less likely to be affected.

Authors would certainly find a change as they would be inputting material directly using a keyboard with editorial and correction facilities immediately available to them. Several people noted that control of the final output was passing towards the author end of the communication chain with the use of electronic means whereas, traditionally, control has been towards the printer end. This, of course, has implications for the work of publishers and printers and it was felt that those involved in specialist jobs in these areas might well view the electronic revolution as rather more of a threat than a promise. This was seen more positively by some as a challenge: 'the publisher must become an information expert'.

'As computers and telecommunications will play a major role in future publishing organisations, specially qualified staff will be required for: library science, information analysis, computer software, etc.'

How should people be thinking and preparing for these changes?

This was seen from two viewpoints, firstly that of the company and secondly that of the individual in the company. It was suggested that the way to move forward was to design and carry out small projects using external expertise where possible. Consideration of the needs of the end user was also stressed.

Introducing changes to one's staff was considered not only in terms of the numerous conferences that abound, but also in terms of the need for in-house conferences for all staff with ample time for discussion where specific problems could be discussed with others sympathetic to the needs of the group.

The future of print on paper

One widely held view is expressed in the comment below: 'Because we already throw away newspapers yet treasure battered paperbacks containing ideas we revere, both electronic and paper borne information will survive the next ten years. Utility information needs only the best utility packaging. A thought of lasting value will for some time be most acceptably owned on that "user-friendly" medium of paper.'

The concept of 'gift' books was mentioned and it was felt that: 'No new technology yet developed is in any danger of displacing the printed page for leisure reading or substantial academic reading. Reasons: convenience, comfort, ease of cross reference, aesthetic satisfaction.'

The future

It was generally agreed that the new technology would displace print on paper for many forms of information but that the future was seen not merely as an alternative between print on paper and 'print' on screen. Rather a multi-media future was seen high quality video and other means of information transfer integrated into a 'home information system'. One guess for ten years hence split this into:
'Print on paper 75 per cent; audio-visual 10 per cent; on-line 15 per cent.

An interesting parallel was drawn between present developments in electronic information systems as compared with paper-based information systems, these being likened to the differences between television and radio broadcasts as compared with the increasing home use of video and audio cassettes.

Television and radio news broadcasts were the audio-visual equivalents of the daily newspaper, and cassetted material could be likened to books that could be used for leisure or learning.

The use of the video screen for the presentation of textual material which is at present usually presented by high quality print on paper was questioned as the video screen is at present associated with high quality colour pictures, sound and only relatively small amounts of text in normal television transmissions. It was felt that video disc technology may bridge the gap and that the future could well be with fully integrated audio-visual systems also capable of presenting high quality graphics and text.

Index